THE HIDDEN FOREST

THE HIDDEN FOREST

The

Biography

of an

Ecosystem

JON R. LUOMA

An Owl Book

Henry Holt and Company

New York

Henry Holt and Company, LLC
Publishers since 1866
115 West 18th Street
New York, New York 10011

Henry Holt® is a registered trademark of
Henry Holt and Company, LLC.

Published in Canada by Fitzhenry & Whiteside Ltd.,
195 Allstate Parkway, Markham, Ontario L3R 4T8.

Library of Congress Cataloging-in-Publication Data
Luoma, Jon R.
The hidden forest: the biography of an
ecosystem / by Jon R. Luoma.
p.   cm.
ISBN 0-8050-6448-6
1. Forest ecology—Oregon—H. J. Andrews Experimental Forest.
2. H. J. Andrews Experimental Forest (Or.)   I. Title.
QH105.07L87   1999   98-46559
577.3'09795—dc21   CIP

Henry Holt books are available for special promotions and
premiums. For details contact: Director, Special Markets.

First published in hardcover in 1999 by
Henry Holt and Company

First Owl Books Edition 2000

Design and frontispiece photo by Michelle McMillian

Printed in the United States of America

1 3 5 7 9 10 8 6 4 2

Acknowledgments

Special thanks to Fred Swanson and Art McKee for hosting my series of visits to the Andrews Forest, and to Jerry Franklin, not least for pointing me toward the project in the first place. Because I chose to highlight only selected aspects of the process of discovery that continues in the Andrews, some of the threads I pursued with various team members could not, unfortunately, be picked up here, and will have to wait for future reports. In the Pacific Northwest, my gratitude, specifically, to Andrew Baker, Lynn Burditt, George Carroll, John Cissel, Warren Cohen, Bill Denison, Gordon Grant, Stan Gregory, Mark Harmon, Julia Jones, Jack Lattin, Dan Luoma, Andy Moldenke, Randy Molina, Nalini Nadkarni, Dave Perry, Nathan Poage, Tim Schowalter, Jim Sedell, Dave Shaw, Phil Sollins, Tom Spies, Jim Trappe, David Wallin, and Carol Wood.

For a general education about new ideas in ecosystem science and conservation biology, many of my sources are in the Great Lakes region. Forest Service ecologist Dave Cleland, in particular, has, with patience and generosity, led me down an intellectual path that, quite literally, has altered how I see nature, in space and time. Key insights into ecosystems, endangered species, and tree physiology have come from Tom Crow, Jim Jordan, John Probst, Mark Coleman, and Richard Dickson, and insights into the realm of conservation biology have come from Bil Alverson, Steve Solheim, and Don Waller.

Thanks, as well, to my editors Bill Strachan and David Sobel and to my agent, Marian Young, who somehow has managed to put up with me through three books and many years. To my wife, Pamela Hendrick, and our son, Ben, love and gratitude for it all. Finally, additional thanks are due Dave Cleland, Art McKee, Fred Swanson, and several members of the Andrews team who read portions of this manuscript for accuracy. Any remaining blunders of fact or context are solely my own.

Preface

Months before I first visited the Andrews Forest in Oregon, I stopped for a visit at the University of Wisconsin Herbarium, the site of a spectacular collection of plant samples from all over the world, and a top center for plant taxonomy. That day, I wanted to pick the brains of a pair of young botanists at the Herbarium and an ecologist they worked closely with, in the hope that they could set me on course to develop an idea that then had only vaguely taken shape.

It seemed like a reasonable and straightforward enough project: I simply wanted to write a book that would help readers, well, be able to see the forest for the trees. A concept fully embraced by then in environmental as well as ecological circles was biodiversity, and in fact the three University of Wisconsin scientists were at the forefront of an attempt to convince administrators from the two Wisconsin national forests to manage their woods more ardently to protect biodiversity. A concept that goes beyond concern for individual species, biodiversity implies concern for life at both a finer and coarser scale: for unique genes within populations of a given species (tigers in Siberia carry a suite of different genes than tigers in India, although they are both the same species), and, on a grander scale, for the communities and ecosystems to which species inevitably, and intimately, must be linked.

From the three Wisconsin scientists—Bil Alverson (most recently on a long-term fellowship at Harvard), Steve Solheim (then a newly minted Ph.D., now a professor in the Wisconsin system), and Don Waller (then and now an eminent ecologist at Madison)—I wanted simply this: some ideas about how to gather material for a book that would describe the interlinkages, the ecological workings, of a naturally functioning forest.

Alverson and Solheim had a treat in store for me when I arrived. I met them in the Herbarium and got a brief tour of this library of plant life, but then we proceeded upstairs to a spacious, high-windowed, sunlit room, where Waller joined us. Once an office, the space had become a conference room, for it was ground too hallowed to dole out to an individual scholar. It once had been the office of one Aldo Leopold, who as a scientist was the father of wildlife ecology in America but, at least as important, was one of this century's great and eloquent environmental philosophers and writers.

Leopold's own conversion to an ecological worldview had come when he was a young man in 1909. Following a practice common then (including in national parks), he found himself one day hunting wolves in the mountain country of New Mexico. He and his companions were eradicating predators to improve nature, or so they thought, making the wild better for game animals such as deer.

That day, Leopold indeed shot a female wolf. He later wrote: "We reached the old wolf in time to watch a fierce green fire dying in her eyes. I realized then, and have known ever since, that there was something new to me in those eyes—something known only to her and to the mountain. I was young then, and full of trigger-itch; I thought that because fewer wolves meant more deer, no wolves would mean a hunter's paradise. But after seeing the green fire die, I sensed that neither the wolf nor the mountain agreed with such a view."

Leopold had gone on to articulate much of what today is entrenched environmental wisdom, including that "all ethics rest upon a single premise: that the individual is a member of a community of interdependent parts."

What better place to begin my quest to understand interdependencies in an ecosystem than Leopold's old office. But that day the three scientists were going to disappoint me. They couldn't be of much help, after all, they said, if I wanted a thorough view of how a natural forest ecosystem really functions. Solheim borrowed from Leopold himself to explain the problem. A great bear of a man who was then dividing his own research time between plant studies in the tropical rain forest and biodiversity studies in Wisconsin forests, he mentioned an oft-quoted Cartesian analogy from Leopold, who had once compared what we now call biodiversity in nature to a watch.

"If the biota, in the course of aeons, has built something we like but do not understand, then who but a fool would discard seemingly useless parts?" he wrote in his essay "Round River." "To keep every cog and wheel is the first precaution of intelligent tinkering."

"But there isn't a forest anywhere," Solheim said that day, "where we even know what the cogs and wheels are!" There is no forest, he said, where scientists have even cataloged all the species present—plants, animals, fungi, bacteria—much less described more than glimmers of the complex relationships between these living cogs and wheels, and the nonliving environment—geology, climate, soil.

More digging, and more questions, showed me just how correct Solheim and company were—or, as Leopold himself had written four decades earlier of the complexity of nature: "Only those who know the most about it can appreciate how little is known about it." It was a baffling paradox. It seems as if we are well on our way to knowing more about the moon and the planets than we do about how critical living systems on Earth truly function.

Several months later, my conundrum unresolved, I found myself in the office of an ecologist at the University of Washington. I had dropped in to interview Jerry F. Franklin for an article about his role in the controversy over the logging of old-growth forests in the Pacific Northwest. I only knew that some environmentalists had dubbed Franklin "the guru of old growth" and that he was then plugging a concept he called the New Forestry. But in the midst of what turned out to be a morning interview that lasted through lunch and into the afternoon, Franklin told me about his long involvement in a remarkable project in Oregon aimed precisely at finding and understanding those cogs and wheels, in a spectacular, wild, ancient woods called the Andrews Experimental Forest.

Two weeks later I was back in the Pacific Northwest, this time for the first of what turned out to be a series of trips to the Andrews, which, as I write these words, is the most studied primal forest ecosystem on this continent, and perhaps the planet. That does not mean that scientists here have found every cog and wheel, much less every relationship between them. But here they have discovered a host of species previously unknown to science, and interactions in the forest ecosystem that no one previously imagined. Here, in the shadows of this woods, in its rivulets and streams, under its soil, and high overhead, they have discovered a hidden forest.

Much of this book is the story of discovery. It is not an attempt to be all-inclusive—too much research has gone on in this place to attempt that, with more than twelve hundred published research reports, articles, and books to date and literally hundreds of others now in the works. Rather, it is a sample of discoveries that seemed to me most engaging, compelling, and precedent setting.

In the end, there is another hidden element in all of this. There are a scattering of other whole ecosystem projects under way elsewhere. But if as many surprises, as much scientific and public policy ferment, as much wonder and pure and practical knowledge about life itself can rise from a single ecosystem study, it begs a question: why in the world are such projects not a top national, or global, priority?

THE HIDDEN FOREST

1

 IN THE HEART OF THE ANDREWS FOREST, in central Oregon, a log lay before us on the forest floor. As a tree, it had stood here from the time of Columbus, growing from seedling, to sapling, budding, pushing out leaves, transmogrifying sunlight to sugar, sopping up water and nutrients from the soil, laying on wood, reaching up, year by year, until it stood nearly one twentieth of a mile high.

Centuries old, perhaps weakened by disease or rot, it had been caught in a fierce sheer of wind only weeks before. Its plummet would have begun in slow motion, gathering speed and power as the ancient, giant bole ripped through the branches of its giant neighbors, finally hammering the forest with a great earthshaking boom.

On this day, early in winter, it was windless, quiet, nearly silent. Fred Swanson led the way onto the log, stepping nimbly up a series of natural, snowy stairs on the log's busted-off trunk end. Swanson, a lean, rangy scientist in blue jeans and hiking boots, wore a full black beard and had eyes so fiercely intense that his friend and colleague, ecologist Jerry Franklin, once suggested that he "looks like an Ayatollah." Lynn Burditt, in layers of shirts under her green wool forest-ranger jacket, with the badge of the U.S. Forest Service emblazoned on it, followed him quickly up the log's stairway. I followed, less nimbly, after.

The log was huge enough that it placed us fully chest high off the forest floor, thus serving as a kind of natural reviewing stand. On this warming December morning, fat wet flakes of snow fell. Before us, all around us, lay a stunning, sterling example of a forest ecosystem that has become famous, or to some minds infamous, in recent years: the legendary old growth of the Pacific Northwest, one of the world's most ancient forests.

For anyone accustomed to the forests nearly anywhere else in the world's temperate zones, the first and most overwhelming impression here is not so much one of oldness, but of sheer immensity. West of the ridgeline of the Cascade Range of Washington and Oregon, the dominant trees in virgin forests like this one are beyond big. Every genus of conifer found in these forests is the largest of its kind on the planet: the tallest spruce (Sitka spruce), the tallest cedar (western red cedar), the tallest hemlock (western hemlock), the tallest pine (sugar pine), the tallest true fir (noble fir), and, most especially, the tree that our impromptu viewing platform had been in life: the Douglas fir, the third tallest of the earth's trees. (This famous fir is not, technically, a "real" fir at all; its scientific name, in fact, is *Pseudotsuga*, Latinized Greek and Japanese meaning "false hemlock." Although it has variously been called everything from Oregon pine to western larch, it in fact belongs to a genus of its own.)

That so many trees evolved as giants here has long remained an unexplained curiosity of science. Comparisons with other ecoregions suggest that neither this region's legendary precipitation, nor a climate moderated by the Pacific Ocean, nor even the rich local soils can fully explain it. Surely these factors are part of the reason. But other geographically similar areas, like the rain-drenched and temperate coasts of England, have never grown such giants. Biologists now believe that these trees adapted immensity as a solution to a problem unique to the Pacific Northwest. As much as 90 percent of the region's abundant rain falls in the winter. In the sunny, dry summers, when the trees most need water for photosynthesis, they face a virtual drought. But by adopting immensity as an evolutionary strategy, the region's trees provided themselves with more intercellular space in which to store water, making the bole (that is, the big vertical main stem of the tree) and the complex of giant limbs of a mature Douglas fir the rough equivalent of a gargantuan hump of a camel.

Whatever the reason, all the immensity made for an odd shift in my sense of perspective. My own boyhood had been spent in the woods of the Great Lakes states and Maine; second growth, mostly, but bits of it grown up since the nineteenth- and early-twentieth-century days when lumberjacks had "slicked it all off," clear-cutting vast acreages, then clear-cutting more, cutting and running westward, imagining a virgin American northwoods that would never end. (But of course, it inevitably would, right here, near the rim of the Pacific.) I'd spent some time, too, in the bits and pieces that remained of intact old forest—or in the presence of a few old trees—in the East. Walking in places like Michigan's Hartwick Pines State Park, or canoeing down northwestern Wisconsin's Brule River, I'd found myself beside some of the continent's largest remaining white pines—towering trees, or so it seemed to me, those few that somehow had been kept away from the lumberjack's axe.

But here, the sheer scale of the trees, standing and fallen, generates the disquieting sense that, standing on our log platform, we are suddenly little Gullivers in a forest Brobdingnag. The trees here are easily three times as tall as the tallest white pines I'd ever seen in the East, and many with trunks fat enough that the three of us, linking hands, could not encircle them. Yet the forest seems proportional to *itself.* Looking out on this woods from the perspective of our log, young trees growing in the understory that would be big, mature trees anywhere else look like saplings here. The saplings growing in patches of light look like shoots, maybe ferns. On old logs, some of them approaching as wide as I am tall, moss grows in mats you could bundle up like an armful of quilts.

Just ahead of us, another tangle of trees that had blown down in the recent windstorm lies on an upslope. In proportion to this gigantic world, it looks like any other jumbled forest windthrow. But Burditt, who is a forest ranger and knows about such things, quickly calculates that each of the downed trees holds about ten thousand board feet of quality lumber: enough in the windthrow to put up a suburban cul-de-sac of middling-size houses; each tree, to a saw-miller, worth something approaching the cost of a new automobile.

For the Pacific Northwest, big stands of old growth like this one long served as the region's economic foundation. One version of logic would have suggested that the foundation had a gaping crack in it: for decades, timber companies were furiously cutting far more giant trees—logging out far more old growth—than nature, and time, were regenerating. In 1981, scientists whose research was based here at the Andrews were suggesting that only about 5 percent of the original old-growth forest landscape lay within national parks or other protected reserves. As for any unprotected old growth, they added, "the end of the unreserved old-growth forest is in sight."

No problem, suggested experts who called themselves scientific foresters. In fact, many were urging that very process along. Their version of logic went like this: giant trees might indeed be dense with high-quality, close-grained wood, but any rational calculation proved that old-growth forests like this one ultimately were "inefficient" systems.

Timber, they suggested, could be grown faster, better, smarter: the old growth could be removed, sawed into fine, straight lumber or spun out into the veneers that would be glued together to become quality plywood. Meanwhile, where the old forest once grew, an improved, far more efficient forest could be engineered into being—on a rich, wet forest landscape that, overcoming decades or centuries, would produce wood fiber for lumber and pulp like a veritable factory.

"I'm clearcutting to save the forest," declared an enthusiastic "logger" in an early 1970s print advertisement from the American Forest Institute. "If I don't clearcut, Nature will—with winds or disastrous fires that burn out thousands of acres."

In fact, as recently as the 1980s, the term *old growth* was little known outside the forestry profession. Within the profession, *old growth* was close to a derogatory term. Stands like the one Swanson and Burditt and I were looking at were commonly called cellulose cemeteries. Some foresters still call stands like this one decadent, with all the overtones of Sodom and Gomorrah that might resonate from such a word. By contrast, a young, neatly organized forest is called thrifty, in the trade.

Indeed, our view from the log was, in some ways, every bit of the model of the primeval woods that alarmed early Americans. "Hideous and desolate," the Pilgrim William Bradford called the New England virgin forests. "Savage and dreary," Henry David Thoreau called the seeming chaos of the mid-nineteenth-century old-growth Maine woods, full of "decaying wood and the spongy mosses which feed on it." One constant sensation in the ancient forest is indeed of disorder, and of rot. Busted snags, dead remnants of huge trees lance upward, here and there, debarked, sun-bleached, rotting on the stump, and riddled with holes—ranging from tiny holes that make parts of the decaying bole look like it was used for shotgun target-practice to big, oblong holes that look like someone took a heavy chisel to the tree. Even without the recent blowdown and other logs, the evidence of decades of fallen trees lies everywhere on the humpy land: moss-covered ridges that once were logs, steep hummocks and dips fashioned by the tipped-up roots of long-ago fallen trees. Of the clearly visible logs, some are fresh, like our log platform. Others have almost completely subsided into the soil. All are in some stage of rot that may have begun here in Oregon a century before Thoreau even went to the Maine woods.

And yet this is no desert. For all the decay, life abounds here. Aside from growing the greatest assemblages of large trees, the Pacific Northwest forests, so droughty but hospitably warm in the summer, so rain-drenched and still generally temperate in the winter, hold a greater mass of living cell matter per acre than even the lush rain forests of the tropics, typically approaching five hundred tons of living tissue on every acre. Lichens and mosses blanket and drape the boughs of the trees. Those holes in the dead snags are more clues to life: small holes made by beetles and other insects that thrive in the presence of rot and death, large holes made by woodpeckers that have tapped or, in the case of big pileated woodpeckers, jackhammered away hard, in search of those very insects. High up above, by night, flying squirrels leap between the limbs of the old living trees; down below in the soil red-backed voles burrow for truffles in the rotting logs; up above, again, their taxonomic relative, the California red vole, scurries high in the old-growth canopy (where it, and its descendants, will live out their entire life cycles, at altitude). Most famously, perhaps, a certain rare owl with a mottled, spotted breast swoops among the boughs, in pursuit of those very rodents, or other prey.

Thoreau may have shuddered, at first, to see it. But an ancient woods like this one is not so dreary after all—it is neither dark nor gloomy: that character actually belongs more to a much younger forest of even-aged trees whose canopy has interdigitated so thoroughly as to lock out light. Here, young trees grow lush in expansive gaps lanced with sunlight, gaps made when an ancient tree like our platform came crashing to earth. Often, rows of little trees, like soldiers, shoot directly out of the wet and nutrient-rich "nurse logs" that the fallen, rotting trees have become.

The ancient Douglas fir forest we looked out upon that day was typical of the remnants of true old growth that survive here in the Pacific Northwest. But there was something more special about this particular bit of ancient woods. It lay within the Andrews Experimental Forest, a sequestered sixteen-thousand-acre parcel of the vast Willamette National Forest (pronounced, in these parts, Will-AM-et) in the central Oregon Cascades, about a forty-five-mile drive from the town of Eugene. In 1948, the U.S. Forest Service had set this parcel aside as the H. J. Andrews Experimental Forest, part of a network of forests intended to serve as living laboratories for studies by the Service's scientific research branch.

Almost certainly, no one could have imagined then the turmoil that would visit the forests of America and, particularly, the Pacific Northwest: battles over wilderness in the 1960s and 1970s, battles over the survival of that mottled-breasted owl—the spotted owl—beginning in the 1980s, and, eventually, battles to preserve the once-disdained old growth itself, this "decadent" ancient forest, by the 1990s.

Certainly, no one—not even the scientists first assigned by the

Forest Service to conduct studies in the Andrews—could have imagined then the turmoil, and the rocking changes, that would visit forest science over the same period of time. In particular, beginning in the 1960s, the science called ecology, the branch of biology that investigates relationships between all the bits and pieces of natural systems, would begin both invigorating and challenging conventional wisdom about what forests were and how they worked. Anxieties among ecologists about the potential loss of species dependent on old growth began to mount in the 1970s. By the 1980s, some ecologists were even beginning to suggest that there might be great value in the old growth that went beyond even the protection of such species. A surge of new, and alarming, discoveries about the way species survived, or failed to, when reduced to small groups in isolated or fragmented habitats spawned a new branch of biology (called conservation biology and, alternatively, the biology of scarcity) and began to change the way science looked at the alteration of wild habitats through, say, logging. And along the way, science began looking at not just old growth, but forests in general, in a new way.

The Andrews Experimental Forest remains virtually unknown among most environmentalists who have fought recent battles to protect forests in general, and the ancient woods in particular. Yet the Andrews has been the wellspring of key discoveries that often led to much of the turmoil, both environmental and scientific.

And if scientists like Fred Swanson and resource managers like Lynn Burditt are correct, it might also be a place pointing toward a new approach—a new sort of ecoforestry, perhaps—that they believe could allow a nation to protect wild forests and have some lumber too.

That morning as we stood on the log, Fred Swanson himself spoke of a paradox: the culture of science creates what he calls a "centripetal force," a force that tugs at scientists and nearly compels them to study not the complexity of large systems, but smaller isolated bits, "the scientist," as he characterizes it, "who is the world's expert on a species of bug that lives only on the north side of boulders in a stream."

Indeed in 1971, the ecologist and environmental gadfly Barry

Commoner had blasted his fellow scientists for just such rigid reductionism. He wrote: "Modern biological research is dominated by the conviction that the most fruitful way to understand life is to discover a specific molecular event that can be identified as 'the mechanism' of a particular biological process. The complexities of soil biology or the delicate balance of the nitrogen cycle in a river, which are not reducible to simple molecular mechanisms, are now often regarded as uninteresting relics of some ancient craft."

There were practical reasons for that. Funding for research, and the tenure and promotion that come with it in the publish-or-perish world of academic science, provides much of the centripetal tug: funding is almost always for short-term studies of, at most, two or three years, with the expectation that the study will be fine enough in scope to yield useful, publishable results in that short time frame. Thus, even scientists who are trained as ecologists are tugged toward the small, the easily definable, the easily quantifiable, the *non*complex, the reductive. Swanson calls it physics envy.

In 1993, not long after our visit to the log, British ecologist Robert M. May reviewed a cluster of studies that analyzed that very issue. He pointed out that one analysis of 308 ecological studies showed that the average length of each bit of research was only 2.5 years. Another analysis of 749 studies published in the prestigious journal *Ecology* showed that a scant 13 lasted even for five years, and fully 40 percent lasted for less than a year. In terms of geographical scale, the problem was even more striking. An analysis of 97 field experiments found that fully 44 percent covered an area of less than one square meter—about the size of a coffee table, with fully three fourths of them falling under ten square meters.

Any project aimed at finding and characterizing all the large and small cogs and wheels of, say, a forest ecosystem would take many years. It would yield results only slowly, at least at first. It would require scientists trained in a reductive world, where narrow specialization was rewarded, to coordinate and synthesize their studies with other specialists—nearly impossible, given the culture of science.

And yet I had found my way to the Andrews because, virtually as Barry Commoner was writing his critique, this old-growth forest was about to become the site of the most detailed effort in history to accomplish precisely what the centripetal force, and all that reductionism, tugged against. Here, since 1970, an interdisciplinary team of scientists has been engaged in a sweeping, continuous effort to study the workings of a whole, natural, largely untouched forest ecosystem: from the soaring trees, to the myriad organisms in the forest soil, to the fungus rotting an old log, and perhaps one day even to the bugs on the north sides of boulders in streams, all the while looking for connections, at least attempting to work across their diverse disciplines, and all the while almost reveling in complexity.

In the post–World War II days when the Andrews was established as a research forest, the Industrial Age was at its zenith. "Science" here, as at dozens of similar experimental forests scattered around the national forests of the nation, focused on one key goal: finding better ways to use forests to produce faster-growing or more disease- or drought-resistant trees, to produce wood products more efficiently: if the forest in general was a factory, the Andrews was to be a key industrial laboratory.

It has become something else. The particular spot that Swanson, Burditt, and I were looking at that day is called Reference Stand Number Two. It is one of dozens of research sites scattered throughout the Andrews. Here, every tree bears a little numbered tag, as does every log on the forest floor. Elsewhere in the Andrews, instruments monitor nitrogen deposits from the atmosphere, the flow of water in small creeks, the flow of water under the soil, the progress of nutrients in the soil, the depth of the snowpack in winter, and even the ponderous movements of parcels of terrain. (In some cases, it is no longer even necessary for scientists to travel from their laboratories in Corvallis to the Andrews. They can simply click onto a World Wide Web page to find out if it's raining in the Andrews, or if much snow has melted on the mountaintops on a spring afternoon.) Automated thermometers keep track of fluctuations in microclimates. (The climate, after all, in the shade of a tree on a north slope is a different matter than the climate in a small nearby gap in the forest.) Slow-action movie cameras periodically snap their shutters to record the arrangement and rearrangement of huge logs, boulders, and other debris in creeks, and the creeks' own overflow of riparian banks in wet seasons, and their constrictions in dry.

In the heart of the summer research season, scientists can be found burrowing in the soil under logs; or trapping insects fifteen stories or more up in the tree canopy with the aid of rock-climbing gear; or scrambling crablike in a neoprene wet suit in a rushing, buffeting mountain stream, sometimes bathed not only in water, but in the field of an electrical current powerful enough to make one's teeth tingle, to look closely at aquatic life. (The electric current stuns fish and amphibians and floats them briefly out of hiding spots.) One scientist routinely chainsaws disks called cookies from old decaying logs to study, painstakingly, the first stages of the long process of rot. (This latter project is unusually optimistic: it will take two hundred years or more and involve generations of scientists slicing off more cookies before the most-studied rotting logs in history become dust.)

A quarter century after the studies began here, botanist Steve Solheim's observation stands: not a single forest ecosystem exists where biologists have managed even to catalog the full panoply of organisms, particularly the thousands of types of insects, bacteria, and other tiny organisms, much less document the complex relationships among and between them. But here, in the Andrews, researchers have in the past quarter century learned more about how a single forest type functions than anywhere else on earth.

In the process of their two-decade-plus study here, scientists working in the Andrews have made a series of astonishing discoveries about the ways that living and dead and never-alive bits and pieces of nature in the forest interact, from the immense trees to tiny mites in the soil. Given how little science knew to begin with, surprises were probably inevitable. But they have been many.

Andrews scientists have discovered an explosive diversity of life-forms, including literally thousands of tiny animal species heretofore unknown to science, so myriad that it will take decades even to finish describing them all and giving them scientific names—much less to determine just what they do in the ecosystem. Another surprise, the discovery of the importance to the old-growth forest of a secret world in miniature called the mycorrhizosphere, where life thrives thanks to a peculiar symbiosis between giant trees and tiny fungi, a dark, moist, near-microscopic world fed by as much as half the energy captured and packaged into sugars and starches by an immense tree.

Still another surprise: even the seemingly useless, and even the lifeless, can matter immensely in the life of an ecosystem. For instance, foresters have long assumed that fallen logs play no important role in an ecosystem. Logs, mostly dead wood, are very poor in most basic nutrients. So it was easy to assume that their decay contributed little in terms of natural fertility to the soil. But the Andrews scientists have found that the logs are nevertheless as important, perhaps even *more* important, to the ecosystem in death as they were in life. (They have found, in fact, that a rotting log contains a greater sheer mass of living tissue than the giant living tree ever did!)

In short, the scientists working in the Andrews have probably come closer than any other scientific group to understanding the fantastically complex web of life in a forest. In the process, they have managed to turn on their heads many of the comfortable, convenient, conventional assumptions about the very nature of the forest and how we should be "managing" it. Ironically, or perhaps inevitably, the Andrews scientists have also sometimes found themselves disparaged by both forest industrialists and environmentalists as a consequence of what they say they have learned here.

Fred Swanson, who began working here as one of the Andrews team's core scientists not long out of graduate school, is now the leader of that team of scientists. For Swanson, and at least symbolically for the team, a defining moment in the process of discovery about the workings of the living forest came not amid thriving life but, consummately, on a landscape of devastation and death. It happened about ten years into the long Andrews study.

"It was an afternoon," says Swanson, "that ranks right up there with the birth of my children. It was just so amazing and interesting."

On May 28, 1980, Swanson would find himself one hundred fifty miles north of the Andrews, crammed into a yammering helicopter with three other scientists headed for the smoldering, ash-blanketed slopes of Mount St. Helens, the Cascades mountain volcano that had erupted only days before.

The Mount St. Helens eruption had been a blast of holocaust proportion. On that afternoon, only ten days after the blast, Swanson was looking down at a landscape barren, an eerie otherworld, turned white with ash, littered with giant, ragged, whitened stumps and ash-covered trees slammed to earth, facing away from the blast crater. The danger was hardly over. Indeed, the volcano was still active enough that spatters of mud balls made of rainwater and volcanic ash were slamming into the helicopter's windscreen. (The danger is omnipresent for volcanologists. One of the group aboard that day was U.S. Geological Survey scientist Harry Glicken, who, Swanson says, was distressed, stunned. Late on May 17, Glicken had left his USGS colleague David Johnson at their research campsite on the mountainside for a routine visit to a town below. The next morning, while Glicken was washing his clothes at a Laundromat, the top blew off the mountain, and Johnson was killed in the blast. (Eleven years later, Glicken himself would perish on the slopes of a Japanese volcano, also caught in a sudden eruption.)

The helicopter landed in a furious cloud of dust stirred by its own rotors. The team climbed out of the helicopter and into a cyclone of ash. Like astronauts on a dusty new moon, they began to scramble along the mountain slopes.

"The guys I was with were all interested in the big landscape," says Swanson. "They were trying to figure out if the landslide from the eruption was so strong it had run right up and over a one-thousand-foot-high ridge. We were all helping out, digging little holes." Digging the holes would allow the scientists to look at differences in layers of volcanic deposits. Deposits from the landslide would presumably be variable, with dirt grains and pebbles and rocks. Indeed they found that evidence, beneath the more uniform, pebbly deposits from the volcanic blast itself.

Peering into one of those little holes, Swanson saw something that perplexed him. Although he didn't suspect it at the time, it was something with dramatic implications for the directions his team would soon take—something, even, with implications for the management of forests of the future. It was not the stuff of that "big picture" of geology. It was the stuff of a very small, hidden picture involving a near-microscopic life-form that already, not much more than a week after the cataclysmic blast, was laying the ecological foundation for a new ecosystem.

"There were these little spiderweb-like threads in the holes," he says, "nearly invisible." In fact, he could really only see the translucent, tiny threads because tiny bits of feathery, near-weightless volcanic ash were hanging from the pit walls on the threads and fluttering in the wind.

At the time, Swanson didn't know what the threads were. But he would learn. Here in the blast zone, in the most inhospitable environment imaginable, those tiny threads had spread themselves through the ash, the future soil of the mountain. They had spread themselves as a living, undersoil web in just the ten days since the eruption. "The mycelia of burn-site fungi," Swanson called them, "the first biological response."

It seemed that the volcanic blast had killed everything in its path, most notably the biggest organisms, from Douglas fir trees to elk. But Swanson would later learn that its heat had, remarkably, also stimulated the growth of others—"some very ecologically important small ones."

Swanson, of course, could not have known that the tiny threads would one day help lead his team to a new way of imagining what a forest can be.

2

THE QUESTION THE TEAM OF SCIENTISTS has attempted to answer in the Andrews is a grand one: how does an entire ecosystem work? Although they may have gotten farther in answering that question than anyone before, they are hardly the first to try to understand and illuminate the tapestry of an intricate and interconnected nature.

The word *ecology* seems often to be confused with environmentalism. But although studying what Aldo Leopold called a wounded world has indeed driven some professional ecologists to activism, the science of ecology itself is a branch of biology focused on studying connections and relationships in the natural world. It is the science of who eats whom, and who lives where, of how energy and nutrients course through the tissue of the living world, the science of the study of the production of living tissue and its consumption and decay. It is the great integrative biological science of not just molecules, not just cells, not just whole organisms, but of the complex dance of nature across space and time.

Surely even prehistoric humans had deep inklings of some sort of ecology, and presumably a profound knowledge that there were relationships in nature, that the migrations of herds, or the migrations of birds, flowed with cycles of climate or daylight; that vegetation changed from one area to the next, that the game or edible plants they depended on, or simply the animals or plants they saw, survived in association with others; that the creatures of a sun-filled meadow were different than the creatures of a deep forest; that the plants and creatures in moist lowlands were not the plants and creatures of the dry upland. Modern conventional wisdom, some of it perhaps romanticized, portrays Native Americans, for instance, as supremely attuned to the land in the centuries before white settlement and industrialization.

In historic times, the philosophers of ancient Greece foreshadowed the work of later natural historians. Plato, in fact, introduced a notion of a great natural balance, suggesting that there existed a flawless order in nature, "the perfect image of the whole of which all animals—both individuals and species—are parts."

But formal, recorded attempts to actually identify just what those parts were and how they interacted would have to wait for the eighteenth century, and particularly for two detailed natural histories. It was no accident that one of those studies came from the son of a village minister in Sweden, the other from a parson in a village in rural England. In the mid-eighteenth century, a surging interest in natural history, in identifying and naming and categorizing organisms, had begun to captivate the intellectual world—and the religious world as well. In fact, both the son of a parson and the parson pursued passionate interests in what today looks much like a pair of whole ecosystem studies. There was more than sheer curiosity here: part of the stated goal of both men was to prove and reinforce the view that God had designed a nature, as Plato had suggested, of perfect balance, perfect harmony.

The Swedish parson's son was Carl von Linné, today more renowned under his Latinized name, Carolus Linnaeus. Linnaeus's most famous work is his *Systema Naturae*, a method of taxonomy that remains the basis for naming and classifying organisms today. We're all familiar with his key innovation, called binomial classification, where an organism is classified first by its genus name, then its species name. (We are genus *Homo*, species *sapiens*. The grizzly bear is from the genus bears, *Ursus*. Today taxonomists often add a subspecies name. The grizzly bear's is deliciously to the point: it is *Ursus arctos horribilis*.)

Before Linnaeus's taxonomy, the field of natural history—the attempt to observe nature in a systematic way—and the predecessor to the science of biology, was drowning in disorganization, for the existing library of living world was a chaos of conflicting common names. It was perhaps only through Linnaeus's development of a coherent system for classifying species that Charles Darwin could, more than a century after Linnaeus, envision the process of evolution, and essentially invent modern biology.

But although he was far more renowned for his taxonomical system, Linnaeus accomplished a lesser-known feat. He penned an essay, "The Oeconomy of Nature," published in 1749, electrifying for its insights, particularly for observations of links between the bits and pieces of nature.

Linnaeus grew up in the Swedish town of Stenbrohault. This was pastoral farm and forest country, where as a boy Linnaeus ranged in the woods and lolled in the fields, "giddy," as he later put it, "at the Creator's magnificent arrangement." Later in life, as a physician and professor, he began to study formally the relationships between plants and animals in nature in the countryside around the village. Linnaeus was hardly looking to build a foundation for a new science. He merely sought to explain the perfection, the intricacy of what he was certain was the divine power's "magnificent arrangement."

But in the process, he anticipated ideas that would be formally defined as ecological processes in the modern era. For instance, he described a function in nature much like the present-day ecologist's idea of niche, which suggests that each species occupies a specific realm, or way of life. To a modern ecologist, a bird species's niche would be its habitat, its nesting site, its foraging location, and what it consumes—essentially, where it lives and what it does for a living. As a general rule of ecology, no two species ever occupy the same niche.

Even if it might seem, for example, that two different types of birds use the same stand of trees for similar foraging, they do not. Research by the mid-twentieth-century ecologist Robert MacArthur provided a classic modern example. MacArthur studied an array of closely related wood warblers that clearly were feeding on insects in the same trees in the conifer forests in New England, seemingly violating the rule. But by simple, close observation MacArthur found that Blackburnian warblers foraged only at the top of the tree canopy, bay-breasted warblers only in the middle canopy, and myrtle (yellow-rumped) warblers only from the lower branches to the ground.

We know today that this is the raw staff of Darwin's natural selection. Natural selection—the survival of the fittest—drives this principle called competitive exclusion. Simplified, it works like this: if a group of yellow-rumped warblers suddenly attempted to feed at the top of the canopy, a battle for resources would ensue, with the only survivor the species best equipped to fill the feeding niche, in this case, almost certainly, the Blackburnian warbler. In "The Oeconomy of Nature," Linnaeus's version of niche was one in which God had assigned each organism its "allotted place" in a world where no creature could "rob those of another kind of its ailment; which, if it happened, would endanger their lives or health."

And Linnaeus, too, outlined the basics of what would one day be familiar to every modern schoolchild as perhaps the most basic concept in ecology, the food chain. "Thus," he wrote, "the tree-louse lives upon the plants. The fly called musca-aphidovora lives upon the tree-louse. The hornet and wasp fly upon the musca-aphidovora. The dragon fly upon the hornet and the wasp fly. The spider upon the dragon fly. The small birds on the spider. And lastly the hawk kind on the small birds."

Linnaeus even documented what he himself called a "succession" in the plant world: in time, he wrote, the wet marsh begins to dry, sphagnum proliferates and, rotting, forms a rich soil, grassy rushes burst into life on that soil, and then, as the former marsh becomes fully dry upland, it becomes "a fine and delightful meadow." This evidence of change, though, was no contradiction to his notion of a timeless balance created by God. His view was that this succession occurred in nature only as part of an endless cycle, much like the water cycle: evaporation, precipitation, evaporation.

His view of what he called the economy of nature clearly was influenced by the fact that the new, sparkling inventions and contrivances of the machine age were permeating, even altering, the entire view of nature in at least the developed world. Organisms, René Descartes had declared nearly a century earlier, were themselves analogous to machines. "From the work of Galileo, Descartes, and Newton, in particular, there emerged the figure of a vast celestial contrivance set in operation by an omniscient mechanic-mathematician," writes science historian Donald Worster. Nature, was a contrivance, a great machine, made of "replaceable parts made to move smoothly together by the external skill of an artisan."

In 1751, only two years after Linnaeus published "The Oeconomy of Nature," Gilbert White, an ardent amateur naturalist, began a nearly four-decade-long study in which he chronicled and cataloged the plants and animals in the countryside around the English village of Selborne, where he was the local pastor. His enormously popular 1789 book, *The Natural History of Selborne*, partially echoed Linnaeus, whose work provided a foundation for White's own study.

"Nature is a great economist," he agreed. White, too, identified what modern ecologists would call a niche. But White went beyond Linnaeus's analysis. He was able, for instance, to highlight how even wastes became part of the great economy. Even the lowliest of the stuff of nature—cow droppings at pondside—he noted, became food for insects and worms, which in turn became food for fish.

Nearly a century later, White's book, its pages often well thumbed, would sit in a prominent place on the bookshelf of Henry David Thoreau, who would delve in detail into the mysteries of the linkages in nature at Walden Pond and in the environs of Concord. At Walden, he would burrow in soil and rotten wood to examine and catalog the tiny creeping creatures within. He would count tree rings on old stumps to try to learn what trees grew around Concord in days gone by and make detailed records of phenology: the timing of the blooming of flowers and the migrations of birds and other seasonal changes. He noted what today would be called symbiosis by an ecologist, in the role that squirrels play in planting oak trees: burying acorns as a future food supply and, fortunately for future oaks, forgetting where some of them were buried. And not a few times, he would rail against those he saw despoiling nature: "If some are prosecuted for abusing children, others deserve to be prosecuted for maltreating the face of nature committed to their care."

His effort to understand and explain nature's links was substantial. Indeed, Thoreau documented, among trees of North America, the same process of succession that Linnaeus had pointed to— the orderly change that natural systems could go through—a kind of community-wide evolution. Asked why pine trees grew up when oaks were cut down, his powers of observation led him to conclude, correctly, that pines could thrive in the full sun of an open field, while oaks could not. Asked why stands of pines eventually became stands of oak, he concluded that although young oaks might not grow well in full sun, they could thrive in the shade provided by pines, eventually to soar to the sky and overtop the pines. Once he even imagined the successional pattern that would obtain if humans were to abandon Concord itself, the kind of pattern of change he had been able to observe, as a surveyor, in old, abandoned fields. First, he suggested, would come "rank garden weeds and grasses in the cultivated land," to be followed by shrubs like huckleberry, chokeberry, and thorn bushes. And then would come what an ecologist, or forester, today would call the pioneer tree species, the trees that thrive in full sun, "the wild cherries, birch, poplar, willows. . . . Finally the pines, hemlock, spruce, larch, shrub oak, oaks, chestnut, beech and walnuts would occupy the site of Concord once more. The apple and perhaps all exotic trees and shrubs and a great part of the indigenous ones named above would have disappeared, and the laurel and yew would to some extent be an underwood here, and perchance the red man once more thread his way through the mossy, swamplike, primitive wood."

Notably, from Thoreau's remarkable lecture "The Succession of

Forest Trees" (presented to the Middlesex Agricultural Society in Concord in 1860) came a question that would eventually come to resonate through the Andrews team's research: "Would it not be well to consult with Nature in the outset? For she is the most extensive and experienced planter of us all." (Some writers have reported, inaccurately, that Thoreau first coined the word *ecology*. He once poorly scribbled the word *geology* in a letter, an error that was misperceived, and thus misprinted as *ecology* in a 1958 compilation of Thoreau's correspondence. The word, in fact, would have to wait for the science to catch up.)

Four years before Thoreau's untimely death at the age of forty-five, in 1862, modern biological science was born. In 1858, Charles Darwin published his earthshaking theory: that species are not made whole by a supreme power, but evolve by what today would be recognized as an ecological process, a process by which species change over time in response to changes, or new opportunities, or new challenges, in their environment. Darwin's detailed answer: evolution through natural selection.

It was only with the theory of evolution in place that ecology could be born as a modern science. And, in fact, it was one of Europe's most prominent advocates of Darwin's theory, the German biologist Ernst Haeckel, who not only pointed out the need for a sweeping integrative science to study relationships and dependencies among species and their environment, but who gave the science its name (or nearly so). Adopting the same Greek root word *oikos*, meaning household, from which *economy* (or Linnaeus's *oeconomy*) had been formed, Haeckel coined the new word *oekologie*. The new discipline, he said, was desperately needed for "the study of all those complex interrelations referred to by Darwin as the condition of the struggle for existence."

Haeckel proposed the new discipline in 1866. But it was not until about the time the Anglicized spelling, *ecology*, was adopted by the International Biological Congress in 1893, that the new science truly began to blossom. In Europe, perhaps the most critical work came from Eugenius Warming, a Danish scientist and scholar. In 1895,

Warming began to lay out in detail the ways in which organisms functioned, not just as individuals, but as parts of interdependent communities. A community's composition, he proposed, was driven by environmental factors, notably soil moisture, but also climate. A natural community was characterized by a property called commensalism, where species were arranged to coexist in such a way as not to compete with one another for resources (again, what would come to be called niche).

In fact, Warming suggested that commensals often provided resources for one another: the tree a nest for birds or squirrels, and shade for herbaceous plants. A bit paradoxically, Warming also concurred with the Darwinian notion that individuals, meanwhile, also were engaging in constant life-or-death pitched battles with members of their *own* species for the resources they needed for their own survival.

The tightest commensal links, he suggested, were purely symbiotic, where the dependence of one organism upon another was total. For Warming, a perfect illustration of such symbiosis was the lichen, the mosslike organism that is really two distinct organisms functionally wedded to each other. One, an alga, photosynthesizes, producing food energy from the sun in the form of carbohydrates. But it is incapable of retrieving such vital nutrients as nitrogen and phosphorus from the environment. Its bound partner is a fungus. The fungus cannot photosynthesize, and so it is incapable of producing its own carbohydrate food. But it is an efficient filcher of nutrients from stuff as seemingly hostile as solid granite. Acting as one organism, alga and fungus feed each other.

Warming's insights would one day be central to the whole-ecosystem forest study at the Andrews. In particular, the Oregon scientists would discover dazzling examples of commensalism and symbiosis. In perhaps the most astonishing case, they would find a tight symbiosis between towering trees, great matted webworks of fungal threads, and a endangered relative of the mouse. They would find, in fact, that some of the world's most spectacular trees soar above the forest thanks largely to the dining habits of tiny rodents. Indeed, the fine, translucent threads Fred Swanson had found in the hole he had dug on Mount St. Helens had everything to do with Warming's thesis.

Warming's breakthrough analysis built a major part of the foundation for modern, scientific ecology. In studying what he called species' communal life, he had asked the question, and provided some tantalizing first answers, about how groups of species function together in space. But another great question begged to be asked, and gradually answered: how groups of species function in time. Darwin had exploded the notion that species were static. Evolution by natural selection meant that new species would develop, others would vanish. But if species changed, that implied that the old Platonic, and Linnaean, ideal of a perfect, unchanging balance of nature had to be wrong. Change, over time, was the very nature of life.

The first scientific attempt to address the question of how the communal life of species changes with time would come almost simultaneously with Warming's studies from the other side of the Atlantic, and from a man who, quite coincidentally, had much in common with the Andrews team's Fred Swanson.

Fred Swanson first hooked up with the Andrews ecosystem project almost casually in its early years. In fact, his academic credentials are not those of an ecologist, or even a life scientist. Swanson's specialty is an arcane one called geomorphology, which is the study of landforms, a sort of stepsister of geology.

As a college undergraduate, he set out to be a pure geologist. He'd been a rock hound as a boy, so the choice seemed natural. But although he completed his degree, much of geology didn't energize him. What did energize him was fieldwork, and particularly fieldwork collaborating with scientists from more than just his discipline. Twice as an undergraduate, he was picked to spend summers at a National Science Foundation–sponsored multidisciplinary field station on the island of Bermuda. There, a diverse group of researchers and graduate students (including the now-eminent evolutionary biologist Stephen Jay Gould) was trying to sort out how life built land: that is, the intricacies of the formation of limestone from the accumulation of millennia of tiny animal skeletons. A later summer job conducting geological field surveys in the Rocky Mountains of Montana, and still later work in Oregon's Coast Range, drew him to graduate work. That led to more fieldwork, including a six-month-long field trip to study landforms in the Galápagos.

In 1972, shortly after Swanson completed his doctorate, came an offer from Forest Service ecologist Jerry Franklin of a postdoctoral assignment, a project that Swanson assumed would last only several months. Franklin asked him to conduct a baseline survey of the geology and landforms of a site east of Eugene, a place called the Andrews Experimental Forest, where Franklin and scientist Dick Waring were leading a project aimed at describing the workings of the forest ecosystem.

In the process of assessing the ecosystem's geology, Swanson walked the ridges, streamsides, and roads of the forest. He admits his project was frustrating. "If you really want to teach someone about geologic mapping, you take them to the Grand Canyon, where the rock is arranged like a layer cake, or to Montana or Wyoming where so much of the rock is exposed," he says. In the Andrews, he says, the underlying geology was almost entirely hidden, under soil, and, most often, under soil that also happened to have giant trees growing on it. But there were hints: a rock outcrop here, the evident layers of rock in a road-cut there. Those bits were pieces in a puzzle, and sorting out the system's underlying geology became a matter of trying to determine how the puzzle pieces were arranged in geologic time. His general conclusion was that the living ecosystem lay over a complex nonliving landscape, formed initially millions of years ago by flows of lava and volcanic debris and "mudflows" of volcanic ash and water, and then shaped dramatically since then by flowing water, landslides, and glaciers. The only exception to that pattern was evidence, here and there, of a layer of volcanic pumice from the massive eruption of the volcano that formed Oregon's Crater Lake sixty-six hundred years ago, about seventy miles to the south.

But Swanson's geology effort turned out to be several interlocked projects, and not so simple or short-lived at all. As he hiked the Andrews, Swanson became curious about what was happening to the

Andrews landforms over time, about why he was finding phenomena like a giant tree tearing in half at the trunk, as if doing splits
on two woody legs, and groups of trees that were leaning, in unison,
downhill like so many towers of Pisa, and deformed trees that had
clearly been leaning at one point in their long lives, but had curled
back upward to the sky. (The boles, or main stems, of trees are
strongly inclined to grow both up to the midday sun, part of what
botanists call a phototropic response, and away from the pull of
gravity, as part of a geotropic response. The effect is still not completely understood, but it is clearly controlled by minute messages
carried in plant hormones that induce parts of the tree to alter the
way in which new cell tissue is laid down.)

Searching the scientific literature, he found that some geologists
had already suggested that the Cascade Mountain region can be subject to a great deal of shifting and moving about, as a consequence of
soft volcanic rock and winter precipitation of rain-forest proportions.
Swanson set about the task of finding out just how much the
Andrews was moving. He found himself measuring and recording
gaps where the stripe down the middle of roads just outside the forest
had formed a zigzag; by determining when the stripe was painted, he
could determine how much the land was moving. He jury-rigged
mechanical stream-flow gauges to measure and record the movement of landforms that seemed to be pulling apart. And he laid out
patterns of, first wooden, and then steel, stakes—forming the stakes
into a perfect square over an area he thought might be moving and
an adjacent area that seemed more stable to determine how much the
square deformed over time. He looked for the initial wound in split
trees, and then simply counted tree rings from the initial wound
to determine how many years it had taken the tree to be pulled that
far apart.

By the mid-1970s, he had found that much of the Andrews's
mountain terrain is in the process of what he named slow landslides.
Ponderous landslides, he might have called them. The slowest of
them, he calculated, are moving at a rate of only a few millimeters, a
tiny fraction of an inch, each year; the fastest slow landslides are
moving at a few feet annually.

"It's all just a great big slow conveyer belt," he says, although he adds, "if you're a tree trying to grow there over several hundred years, it can be clipping along at a pretty good rate."

But slow as they might have been (and not really slow at all to a geologist, trained to think in terms of millennia to eons), the discovery of these landslides was a first strong signal of what would become a recurring theme in studies of these ancient woods: that the only real constant is change, that for all the appearance of stability, the ecosystem is always in flux—albeit change often nearly invisible to a casual observer because its scale overwhelms the usual human idea of what constitutes a long time. His discovery that the forest moves—that many of the towering trees have traveled, sprawling root system and all, several feet during their centuries-long lives, had Swanson hooked on studying not just the land, moving or otherwise, but its relationship with the life, the ecosystem, growing on that land (although he suggests that he turned to studying living and, in the form of giant trees and logs, dead plant tissue because only about 2 percent of the Andrews is exposed rock. "I just surrendered to the phytomass," he says).

Swanson's specialty, geomorphology, is rare enough that in years of reporting about nature and environmental science, he was the first I'd ever encountered. He is, in fact, only the second geomorphologist I'd ever even heard of. Oddly enough, the only other followed a remarkably similar path, and nearly a century before Swanson. After his discovery of the slow landslides, Swanson would find himself drawn farther and farther into studies where the inanimate interacted with the living: into, in fact, the science of ecology. So, too, with the other geomorphologist, whose fascination with a specific landform quickly led him to ecology, too—and in such a major way that some would credit him with founding the science in America.

In the year 1894, Henry Chandler Cowles was riding in a passenger car on the Michigan Central Railroad, completely unaware that he was in the first years of a long journey that would lead him to being called America's first professional ecologist. He was a young man, then only twenty-four, and bound for graduate studies at the recently created University of Chicago. Cowles hardly could have known when the train pulled away from the station at Oberlin, Ohio, that his trip to Chicago was going to be interrupted. Certainly, he never could have predicted that he was about to lay the foundation for a science so new that it barely had a name.

One can imagine Cowles on the train that summer day, the black, smoke-spewing locomotive chugging along the rolling, sandy terrain that lies just south of Lake Michigan in Indiana. Summer temperatures run hot and humid in Indiana in the summer. Perhaps Cowles, in whatever version of Edwardian gentlemen's clothes he wore for travel that day, was less than perfectly comfortable. From his window, this young scientist of landforms saw landforms the likes of which he'd never seen before. He had grown up in Connecticut, and had seen sand dunes on the Atlantic shore. But here were mountains of sand, some of the highest dunes in the world. As the train worked its way through and among the Indiana dunes, Cowles would have seen something else even more puzzling—a variation in vegetation as dramatic as the towering dunes themselves: dune grasses in one spot; scattered savannalike forest in another; deep, dark, mature forests in yet another.

Astonished, and on an impulse, Cowles shot off the train at the town of Miller, hurriedly located a carriage for hire, and rode back to a nearby spot where he could climb over the dunes and work his way to the shore of the virtual freshwater sea that lay beyond. There, he turned his back to the giant lake's rolling surf, away from the water, and he began to walk.

Cowles was moving through more than space. As he would later decide, he was moving backward in time—at least ecological time—accomplishing what would have been impossible on most other landscapes, tracing one of the great themes that would provide shape for the newborn science of ecology.

As he wandered the dunes, Cowles came to realize that this was something spectacular. Here, Cowles sensed, was a sort of time machine. His geomorphologist's knowledge led him to surmise that the dunes, which rolled away from the lake for miles, had been formed beginning with the retreat of the ice-age glaciers. As they melted and receded, the glaciers progressively left behind new, near-barren sandy shores. Yet as he walked, Cowles quickly realized that not only were the dunes just back from the scoured-clean beach not barren—they were covered with the beach grass called marram—but progressively farther back the plant community continued to change, from the grass, to shrubs and herbaceous plants growing on a richer overburden of soil atop the sand, and finally to forest.

Cowles quickly surmised that he was walking, in sequence, from young new dunes near the lake across progressively older dunes farther back, and thus through various stages of ecological time. In fact, his walk that day was only the beginning. At the University of Chicago, he quickly shifted his area of study to botany, under the world-renowned scientist John Coulter, who had himself become convinced that botany could be much more than collecting plants and discovering their names. Coulter had already gotten word of dazzling new work coming from the studies of Eugenius Warming, the pioneering Danish ecologist. Impatient with the dearth of translations of Warming's work, Cowles, who had excelled at Latin and Greek as an undergraduate, simply learned Danish. And again and again, he went back to the dunes, a half hour by train from the university, and trudged some more. Intellectually, his training as a geomorphologist prepared him perfectly for looking at change in ecosystems over time. Geomorphologists, after all, are trained to look at how time and wind and rain and glaciers change the land.

Cowles concluded that the marram grass colonizing a bare dune could survive high wind and searing wind, even burial by sand and direct sunlight. But once it was firmly established, the marram not only stabilized the landform—the dune—it provided for other lifeforms a bit of shelter and cooling shade, and as it lived, then died, its decay made just a trace of nutrient-rich soil. In time, less hardy plants eventually could prosper. Eventually those plants would crowd out the marram, building more soil, more shade, making way for a new guild of species, and on and on until the ecosystem reached a "climax" stage, a mature, dense forest that would operate—or so Cowles supposed—in a steady, stable state. In time, the progressive shade and shelter of plants and the steady decay of accumulating life would change the very character of the soil, and then of the ecosystem itself.

Warming's theories about what organisms lived where, and in association with what other organisms, would come into play in a changing, evolving sequence.

Almost precisely a century after Cowles's legendary walk, Laura Gundrum, a National Park Service naturalist at the Indiana Dunes National Lakeshore, volunteered to retrace Cowles's footsteps with me. Back over the near-barren foredune we trudged, on a protective boardwalk through marram on the dune's windward side, then through sand reed grass and little bluestem on the lee, then down off the back into an interdunal swale where cottonwoods, usually the first pioneering dune trees, grew among plants like bladderwort and wild grape and sandcherry. A few hundred feet more, where those pioneers had built enough soil (though still a poor soil), we hiked through a dark grove of jack pine. Then up another dune, where a disturbance called a blowout had destabilized the dune and had already buried, killed, then partly uncovered a stand of cottonwoods that now stood bleakly, like half-buried skeletons atop. "A ghost forest," Gundrum called it. Here, in an ecosystem still not stabilized, the successional process would have to start its march anew. Now we climbed up the sand to the height of the handrails on the boardwalk, which themselves would soon be subsumed. Then down again into another interdunal swale that itself would probably fill with the residue of the walking blowout. Then finally we found ourselves in a lovely, dark, deep, and mature woods of oak, witch hazel, hop hornbeam, and sassafras.

Cowles's classic paper "Ecological Relationships of the Vegetation of the Sand Dunes of Lake Michigan," published in 1899, stunned biologists around the world, for it established, once and for all, that communities of species move through a sort of community evolution, at each successional stage laying a biological foundation for the next, and simultaneously for their own demise.

Warming, and, for that matter, Darwin, had provided the new science of ecology with one of its great and enduring research topics: the notion that species both competed and cooperated with each other in a host of ways. Cowles had offered another. He had proven that the natural world was not a great, unchanging clockwork. The natural world, his studies suggested, underwent dramatic change over time. Ecology became a science of dynamism.

(As an indication of the effect Cowles's studies had on the scientific world, consider this. In 1913, ten of Europe's leading botanists and ecologists were planning a trip to America as part of a tour of world plant geography called the International Phytogeographic Excursion. All were asked to make a short list of their top must-see ecosystems. The four sites that appeared on all the lists were the Grand Canyon, Yellowstone, Yosemite, and a walk across the dunes of Indiana with Henry Chandler Cowles.)

Succession would become one of the great and enduring topics of ecological study. Understanding the nature of succession is critical to understanding forests, especially forests that might burn, or die back from disease or drought, or be wiped out by a volcano, or felled by the chain saws of an army of loggers. These are forests, as ecologists like to put it, that have been disturbed, ecosystems that now must recover. If any researcher, for instance, were to propose that there might be a better, safer, ecologically healthier way to remove lumber from a forested landscape, a thorough knowledge of how succession really works on that landscape would be the keystone. Certainly if it turned out that science would one day learn that disturbance and recovery were the very nature of a forest's journey through time, understanding succession in detail would be the keystone again.

With its foundation built, the science of ecology would virtually explode into being in the twentieth century. From Frederick Clements, a near contemporary of Cowles, would come a series of forceful studies focused on man's disruptions of nature. Clements, who worked mostly on the ecosystems of the Great Plains, would by the 1930s make a point that might seem obvious today, but was virtually heresy at the time: the horrendous dust bowls of the Great Depression were a direct result of busting the sod and attempting to improve on nature by cultivating row crops on the dry, short-grass prairie of the American West. Clements was suggesting, given climate and soil and moisture, that parched regions had evolved their plant (deeply rooted grasses) and animal (grazers on the grass) associations in a manner that ran directly counter to what the sod busters were trying to do.

In 1933, from scientist Aldo Leopold came the development of a new science of wildlife biology based on ecological principles. But also, from Leopold and others who had found their way to this still-new science called ecology, a series of warning bells began to toll. In 1935, the ecologist Paul Sears warned of desertification in his eloquent book *Deserts on the March*. Fairfield Osborne told the world about *Our Plundered Planet*. And Leopold warned of the dire consequences of dismantling ecosystems wholesale in *A Sand County Almanac* and *Round River*, particularly of the "prodigious achievements of the profit motive in wrecking land."

In the years after World War II, ecology as a science began to go through a technological revolution that began to change its scope and dimension. In the days of Warming and Cowles and Clements, ecology had remained much like the natural history that had preceded it. It was a science that simply described the natural world. But suddenly new tools were appearing. And so was new money, in the form of federal funding and a growing national concern about such problems as the fate of nuclear residues in the food chain. (By the 1960s, the nuclear laboratory complex at Oak Ridge, Tennessee, would also become the largest and best funded ecology laboratory in the nation.)

Physics envy was getting some satisfaction, for ecology was becoming more a science if not of reductive minutia, at least of more rigorously quantifiable fact. Ecologists had at their disposal not only new tools for measuring the world, such as chemical or radioactive tracers that could allow a scientist to follow the flow of, say, sugars through a food chain. They also had new, powerful, and more rigorous mathematical tools from the exploding arena of statistical analysis. And they had access to computers to conduct such studies. More and more, ecology was becoming a science like—well—physics. Beyond observing nature, ecologists were beginning to experiment with nature, manipulating bits and pieces of ecosystems to determine how they responded. Some scientists were, in fact, talking about the rise of a "new ecology."

In the age of mechanization, René Descartes had invented a schema—a new way to view nature. He had suggested that organisms, and nature itself, were machines. By the late 1960s, the rise of the computer had brought a new paradigm to ecology. Leaders in the field had posited that ecosystems were, at least metaphorically, cybernetic. That is, they could be compared to a logic circuit. Ecosystems were driven by a process like a helmsman at the wheel of a ship. The ship and the helmsman, after all, are really an intricate system of feedbacks and corrections consisting of the compass, the helmsman himself, and the wheel connected to the rudder, controlling the ship. The ship begins to veer off course, the compass provides a feedback to that effect, the helmsman makes a correction by turning the wheel, which moves the rudder, and the ship corrects itself to a true course, until it wanders just a bit again. A population of lynx booms. The hungry cats overdevour their available prey, the snowshoe hare, until the hare population plummets. The lynx population follows, plummeting, too. But then the hare population, with little survival pressure from predators, soars, with the plants the hare eats soaring or plummeting in some kind of synchrony.

But how to understand the great intricacies of the ship and the helmsman, of this great logical circuit? The International Biological Programme (IBP), an effort of the United Nations that began in the 1950s, set out to provide an answer. The IBP had been created to provide solutions to a growing global sense of concern about pollution and environmental destruction. But to save nature clearly would require science to understand ecosystems in far more detail. Out of a meeting of top scientists from the American branch of the IBP at Williamstown, Massachusetts, in 1966 came a striking idea, what scientist Robert P. McIntosh would later characterize as "a valiant and unprecedented effort" to organize interdisciplinary teams to conduct "cooperative studies of whole ecosystems."

All that was needed was someone to take up the challenge.

3

ANDERSON HALL SITS, formidable, majestic, on the handsome campus of the University of Washington. A somber stone building, it is high-gabled and high-windowed, and nestled quietly in a grove of tall, dark, and spirelike conifers. On a gray and misty Seattle day, Anderson Hall looks, for all the world, like a church. And maybe, in its history, it has been just that—a sort of cathedral of silvaculture. Here in the Pacific Northwest, this region built by big timber, Anderson Hall is home to the University of Washington's College of Forestry, one of the nation's leading institutions of its kind, where novices have come to learn the rites of plantation, rotation, board foot, and mean annual increment.

In the view of many environmentalists, too many of America's forestry colleges, including this one, have too long preached ecological blasphemy. Some environmentalists insist, too, that schools like this one have graduated too many "sawlog foresters" trained to envision the nation's woodlands not as rich and wild and diverse ecosystems, but as factory-farms for manufacturing logs for pulp and boards. That way of thinking, say the critics, has long dominated not only the ranks of timber industrialists managing their private tree farms, but also the ranks the U.S. Forest Service, the nominal steward of hundreds of millions of acres of America's public woods. Right here at Anderson Hall, the harshest critics have fairly sneered, is the real U of W: the "University of Weyerhauser" (referring to the timber giant Weyerhauser Corporation).

The criticism has deepened in the past decade or so, as environmentalists and timber industrialists have argued with increasing heat over the fate of the giant firs, hemlocks, spruces, and cedars, of spotted owls and marbled murrelets, in the old-growth woodlands that lie within the public boundaries of national forests in Washington and Oregon.

But these days in Anderson Hall, not everyone worships at the altar of conventional forestry. I found Jerry Franklin in his cramped and cluttered office on the second floor. Willingly or not, Franklin, a graying, bespectacled ecologist (whose middle name is literally Forest), has in recent years been at the focus of the debate over old growth and has maintained a position on the cutting edge of the debate about how the woods should be managed (or not).

Franklin was once a scientist with the Forest Service but, when I talked to him, was installed in an endowed chair as the University of Washington's Bloedel Professor of Ecosystem Analysis. And whatever else Jerry Franklin might be, he is not, by any measure, a traditional "sawlog forester."

The Forest Service, his full-time employer until the late 1980s, called him its chief plant ecologist. Some environmentalists once called him "the guru of old growth" (although later many decided they were angry with the guru). In a scribbled inscription on a photograph on Franklin's office wall, Dale Robertson, the former chief of the U.S. Forest Service, jokingly (or perhaps only half-jokingly) needled him for being the service's resident "loose cannon."

But in fact, Franklin spent much of his early career as neither guru nor loose canon but rather, he says, as a conventional, practical forestry scientist with the U.S. Forest Service research branch, conducting studies of how to log and replant and grow forests in ways that provided maximum timber output from the woods.

But Franklin says that during the 1960s, he began wondering

about a new sort of problem, asking questions that would put him squarely in the company of his intellectual ancestors—Linnaeus, White, Cowles. Even as a young boy, he was fascinated by the old-growth forests. But as an adult on his way to becoming one of the nation's leading forest ecologists, Franklin was well aware of the designs the Forest Service had on the region's remaining old growth. "Almost all of it was going to be liquidated," he says, simply. "That was what we all assumed."

Just before the turn of the century, at about the same time that biologists like Eugenius Warming and Henry Cowles were laying the groundwork for the new science called ecology, the concept called scientific forestry was being born in western Europe and being transferred, quickly, to North America. This European-style scientific forestry arrived on the American shore as an antidote to an ongoing catastrophe. The wholesale removal of native forests had begun on this continent with the first white settlers. In the eighteenth century, as the American population boomed, it had accelerated into a devastating rape-and-run style of logging, leaving virtually leveled first the forests of northern New England, then the forests of the Great Lakes states, and then the vast pineries and hardwood stands of the South, as lumbermen slashed and sawed their way westward across the continent.

In their wake they left little but devastation. In what had been the great white pine forests of the Great Lakes states, for instance, they often left huge amounts of unwanted bark and limbs and other "slash" piled on the ground—sometimes as deep as fifteen feet—great mounds of dry fuel. Cataclysmic fires routinely followed. The fires were fueled, too, by blown-down, unwanted "junk" trees loggers had left behind. An estimated fifteen hundred people lost their lives in one firestorm that swept into the logging village of Peshtigo, in northern Wisconsin, in 1871. Witnesses described fireballs exploding in the air, probably superheated gases, and whirling "tornadoes of flame." Another 418 villagers perished in the great Hinckley, Minnesota, fire of 1894, many of them asphyxiated when they retreated into a millpond to escape a fire so fantastically hot that whole trees and large flying limbs went sailing in the firestorm winds overhead.

In the Ernest Hemingway story "The Big Two-Hearted River," the character Nick Adams, himself weary and burned out from the ravages of war, takes the train to go trout fishing, near the northern Michigan town of Seney, just after the area has been visited by such a legendary fire. "There was no town," Hemingway wrote, "nothing but the rails and the burned-over country." In fact, Hemingway had himself gone fishing here as a young man, in 1919, after fires had leveled Seney after logging had cleared all the big pines out. In the story, all that remains of the village is the foundation of a hotel. But most telling is what the fire, in story as well as reality, did to the landscape: "Even the surface had been burned off the ground."

In fact today, if you drive the old, sandy road called the Adams Trail that runs to the northwest of Seney, near the more prosaically named Fox (the river Hemingway's Nick Adams would have really been fishing), you can see evidence of the very worst the rape-and-run loggers ever wrought as they relentlessly worked their way west. The road passes through the vast, flat terrain called the Kingston Plains. It is instantly evident what this place used to be like. Everywhere there are immense, charred stumps of the great white pines that once towered here. But since "even the surface had been burned off the ground" in the hot fires, the Kingston Plains today are little but sparse grass and scrub growing on an impoverished, still almost nonexistent soil.

Bernhard E. Fernow, a German-born and -trained forester who served as chief of the agency that was the predecessor of today's Forest Service, was a key figure in the introduction of the science of forestry to the United States. As early as 1895, he was explaining to American students, politicians, and businessmen its core principle, that good forestry "is exactly the same as agriculture. It is the application of superior knowledge and skill to produce a wood crop." According to scientific forestry, the woods are not simply there for the raping. Their resources need to be husbanded. To cut a forest means to plant a new forest, and nurture it to maturity—to be cut again. Fernow's scientific forestry flag would, by the turn of the century, be waved most vigorously by Gifford Pinchot, the European-trained American forester who succeeded Fernow as chief of the U.S. Department of Agriculture Division of Forestry. Pinchot, who would become the Forest Service's first chief, as it became a separate branch of the Department of Agriculture (under whose administrative wing it remains), also clearly saw forestry as little more than a form of farming.

"Trees may be grown as a crop just as corn may be grown as a crop . . . the farmer gets crop after crop of corn . . . the forester gets crop after crop of trees."

The concept today might seem evident, almost foolishly simple. But at the end of the nineteenth century, in a nation where the forests had once seemed endless, and woodlands had been treated as virtual mines full of logs, it was a startling notion. But by the early years of the twentieth century, the realities of destructive forestry were evident to anyone with eyes. Rape and run had done its worst to the American forest. The last remaining wild woodlands lay only in the West, the best of that in the Pacific Northwest. Beyond, there was nowhere to run, for beyond lay nothing but the Pacific Ocean. By 1909, the manager of Weyerhauser Timber Company, one George S. Long, was offering firm support to this new, even radical idea. "Timber is a crop," he stated flatly. Still, it took until the 1940s for the agricultural model of Fernow and Pinchot's "scientific" forestry to take solid hold in the region.

Whatever advantages it offered in terms of regenerating woodlands, the precepts of scientific forestry never extended to husbanding natural forests, and certainly not ancient natural forests, any more than "scientific" agriculture proposed feeding the world or generating profits from a wild prairie. In fact, as those American foresters who came after Pinchot saw it, the ideal scientific treatment for a truly wild forest, and certainly for an old-growth forest, was obliteration in the name of efficiency.

An ancient forest stand, like the one Swanson and Burditt had led me to in the Andrews, contains a diversity of trees, shrubs, and herbaceous plants. It contains, often, another kind of diversity—a great diversity of ages of living trees—with younger trees growing in the clearings where old trees have died and crashed to earth, or in openings made by burns or windthrown stands. It contains, perhaps most striking to the eye, a diversity of structure—standing live trees in an array of sizes, sapling to cathedral giant, along with those rotting snags, dead but still standing—and those enormous tons of logs.

But a virtual commandment of conventional forestry since its birth early in this century in central Europe has been that younger is better than older, and consistency is better than diversity. In the early years of their lives, plantation trees in a full bath of sunlight can grow vigorously, photosynthesizing the sunlight along with carbon from the atmosphere into sugars, and processing the water and nutrients from the soil into cell tissue, adding thick annual rings of wood fiber around their boles.

Acre by acre, such young, growing stands of trees can add astonishing amounts of wood fiber each year. In about eighty years, solely by seizing the nutrients in the soil, carbon dioxide in the air, and the photon energy in sunlight, a newly planted thousand-acre stand of Douglas firs can grow from virtually no wood, to five million board feet of lumber—enough wood to build five thousand American houses.

But the growth does not continue forever. Although trees themselves will add wood as long as they continue to live, the rate of growth normally slows down. Worse, from a lumber-production point of view, old stands of trees can stop adding new wood at all, because even while some are growing, others are rotting and dying. Ancient forests have reached a sort of biological steady state. In a steady-state old-growth stand, many of the trees are approaching the end of their long lives. Although their wood is superbly clear, strong, and, hence, valuable, it is also more susceptible to rot as long as the tree remains in the forest. As old trees die and crash to earth, new young trees seed and grow and even thrive in the opening, adding back living mass. Still, in the old growth, acre for acre, the forest typically neither loses nor gains any wood. From a wood-production perspective, this steady state is like a savings bond that has ceased to accumulate interest. For this reason, conventional foresters have long viewed old-growth forests as "inefficient."

Even a young forest not logged under a careful management scheme can be considered inefficient, according to this paradigm. A newly planted forest's growth rate will rise every year until it reaches a sort of plateau, with growth leveling. Foresters call the plateau, where yearly returns flatten, the "culmination of mean annual increment" or "economic maturity."

Most revealing is that, in the jargon of conventional forestry, a forest that has still not reached the peak of its annual mean increment is called "thrifty." Economic maturity can come decades, even centuries, before the tree would be classified as mature, and even longer before it would be called old growth. On the other hand, should a succession of foresters allow an expanse of woodland to mature past the peak in annual mean increment—say, for two hundred years— or long enough to begin reaching old-growth status, the forest would be the opposite of thrifty. The term for that, still in use today among some foresters, is *decadent*.

To turn a forest into an excellent tree farm demanded leveling the chaos of the old growth. Once the unruly natural forest was cleared, foresters could scientifically select the best species to grow within the new woods. In the Black Forest of Fernow's native Germany, that might be Norway spruce or silver fir. Here in the Pacific Northwest, the tree of choice usually was Douglas fir, the towering plant that loggers and foresters like to call "the money tree." The quality of its wood, and its economic value as softwood lumber for house building and other construction, was, and remains, extraordinarily high: it is easy to cut in all directions, unusually strong for a softwood, even-grained, not inclined to warp. Douglas fir two-by-fours form the hidden framework of millions of American homes, its strong veneers the layers of plywood that sheathe them. One 1990 estimate suggested that about one fifth of the total volume of timber harvested in all of the United States came from this single species of tree, which grows commercially solely in the Northwest. And for the prized Douglas fir, scientific management offered another commercial advantage.

As Cowles and Clements had correctly documented, ecosystems can be dynamic, moving through a gradual process of succession. Despite their enormous size, and their ability to live enormously long times (the oldest known fir was over thirteen hundred years old), Douglas firs are not the "climax" species of their forests. That distinction belongs to other, less valued species, like western hemlock and western red cedar, which would grow happily in the shade, eventually overtopping and shading out of existence the firs. But by following an agricultural model, foresters could easily plant and grow only firs, harvesting them, and restarting the succession process, long before the climax trees could grow up and impair the vigorous growth of more valuable firs—and in fact eventually shade them to death.

Once the woods were converted to a log farm, lumbermen could move like a threshing machine through the managed monoculture of trees as if it was giant wheat field, periodically harvesting the logs in great clear-cuts. Although in Fernow and Pinchot's day logging still meant axes and saws and teams of horses, within a few decades full mechanization would arrive, in the form of howling chain saws and immense log-moving machines. (In forests with more modest-sized trees, logging is now accomplished with a rig called a buncher-feller, a sort of cross between a bulldozer and a giant robotized scissors that eliminates the need for saws at all by simply snipping off, at tremendous force, the trunks of trees as if they were blades of grass.) Scientific foresters could supervise as the site was regenerated, often burned to clean out any debris and slash, and, since the chemical revolution following World War II, often dosed with powerful herbicides to suppress the growth of shrub and fern species that might compete for sun and moisture with newly planted young firs.

With the terrain on the site thus nicely processed and prepared, the new seedlings could be planted, perhaps even in neatly arranged, arrow-straight rows. Bob Spense, a lumber milling company president from Seattle, once told me, "We can grow trees just like corn." And he had shown me, flying me out of the Seattle airport deep into the Cascade Mountains in Washington in the corporate helicopter he uses to reach his company's mills scattered throughout the region.

The "just like corn" trees were growing, apparently robustly, in precise rows on corporate land below. Later, when we went yammering over the Gifford Pinchot National Forest, not far from Mount St. Helens, he'd pointed, with a wince, to old growth. "If we don't take those old trees out of there," he said, "that so-called ancient forest is just going to fall down."

By 1969, given the locked-in conventional wisdom about forests, no well-behaved scientific forester—or even a forestry scientist with some exposure to ecology—should have dreamed of suggesting that maybe, just maybe, there were unrecorded, undiscovered, but substantial values in the decadent old growth itself.

Jerry Franklin recalls of the late 1960s: "There certainly didn't seem to be any reason to study old growth. Forest researchers had assumed that these forests didn't really have any value, since they're not really producing any additional boards as the years go by. They're just sort of sitting there. We all grew up with terms [for old growth forests] like 'biological deserts.' Some people called the old growth a 'cellulose cemetery.' "

Ideas like those of Spense, he told me, reigned supreme. "There really was a sense that if we didn't get out there and log out those decadent forests, in twenty or thirty years all the trees were going to fall down and there wasn't going to be anything left to pick up. Everybody believed the idea that you're taking care of the land if you can get a good stand of new trees growing after you cut all the old trees down."

That, he would one day come to conclude, is also at the heart of the problem with modern forestry, a conclusion that began with visits to the Andrews Forest.

For the Andrews in Oregon, just like other U.S. Forest Service experimental forests, that conventional wisdom dictated that research *should* be focused on regenerating young trees. And that, says Jerry Forest Franklin, is precisely the sort of science he practiced as a young Forest Service researcher in the early 1960s. Already, he and other forest scientists were supervising patchy clear-cuts and other logging activities throughout the sixteen-thousand-acre

Andrews site, experimenting with different approaches to boost regeneration.

"Basically," he says now, "I had the same assumptions everybody else had coming out of forestry school. Now I look back and think about how dumb I was. I look back and I say, Judas Priest, how could I have not appreciated all the things that *must* have been going on in these ecosystems? I like to tell students now that in another ten years, you're going to look back and say, 'I can't believe nobody realized this, whatever *this* is.' "

Still, Franklin says that since a boyhood in a pulp mill town in southern Washington, he had been fascinated by wild old woods, where he had often gone camping with his family. And now, in the Andrews, Franklin kept finding himself under the forest giants of the old woods. He was acutely aware that although entire landscapes like these once swept across the nation, no one had ever made an attempt to study an old-growth forest in detail.

In the late 1960s, he says, "I got this idea that it might make sense to study these forests in more detail before they were all gone. People in the profession thought I was off my rocker. The attitude was completely pervasive. People said, 'What in hell do you care about these old forests for? It's obvious that the future is in the young, managed forests.' " Franklin's notion seemed to be doomed.

Then in 1969, Franklin's chance came when the National Science Foundation offered extensive funding for studies of forests in the region as part of the new International Biological Programme. Ironically, Franklin's present employer, the University of Washington forestry school, nearly beat him out of his chance to explore the old growth. The Washington forestry school researchers wanted funds to go toward studying the ecology of younger, biologically simpler, plantation-style forests, the thrifty forests of the future. But somehow Franklin convinced the foundation to split the funding, allocating about half to Oregon State University for his unconventional old-growth ecosystem study.

Franklin had proposed nothing short of an all-out frontal assault on the knowledge vacuum about old growth: that a team of scientists from an array of disciplines would combine their skills to begin the long process of understanding just what made an old-growth ecosystem tick; that a small army of specialists from both the federal agency and from Oregon's two major universities, including botanists, entomologists, mycologists, mammalogists, biochemists, ornithologists, plant and animal ecologists, hydrologists, and others descend on the Andrews Forest.

And descend they did.

It took time, well over a decade, for some of the most critical discoveries here to begin to converge. But in the end, the skeptics would have been proven profoundly wrong. Most of all, here in the Andrews Forest, the long years of research that Jerry Franklin imagined into being led to the discovery of a host of previously little-known ecological processes that would begin to call into question much of what scientific forestry thought it knew about the woods.

4

WE WERE FLYING over the roof of the forest. At least it seemed like flying—a barnstorming, open-air kind of flight, in the chilly air of a January day. We were just above the canopy of the ancient forest. Beyond, low mountains rose all around us, old extinct volcanoes, capped with snow and shawled in a maundering gauze of cloud.

The deep-green treetops below us were staggered, some bursting sunward high above others, a jumble of steps, like a sort of green skyline. Indeed, Jerry Franklin compares the view from above the treetops here to looking down on Manhattan. The roof of an old-growth forest is far more dramatic than one could guess from the ground, and the gaps between these giant skyscraper trees are far more evident from above than below. (Looking up from the ground, one sees a more uniform ceiling of green because the limbs and leaves of shorter, subcanopy shrubs and trees fill in those gaps between the old giants.)

We certainly might have been at the level of a Manhattan rooftop, for beneath us, the trunks of trees reached to the ground fully twenty-five stories. Suddenly, at this dizzying height, we stopped. Or more properly, crane operator Mark Creighton, who was sitting it a small glass booth 250 feet in the air, stopped us. He was also sitting about a football field away from us, directly down the length of the massive steel boom that had swung us over the forest canopy. Two years earlier Creighton had actually been building skyscrapers, running giant construction cranes like this one. (Indeed, only three years before my visit, this very crane was topping off a skyscraper in downtown San Francisco.) But he has become a University of Washington employee, certainly the only skyscraper construction expert employed by a university forestry school. Now his crane sits, anchored in three hundred cubic yards of concrete, virtually invisible until you come upon it in the forest. It sits square in the midst of a four-century-old virgin stand in the Wind River Experimental Forest, just north of the Oregon-Washington border, a sister laboratory forest to the Andrews.

We were enjoying a magnificent ride, although we already had been up here in the open air for a half hour and our toes were chilly. There were three of us, wearing hard hats and strapped into safety harnesses clipped to the sides of the rectangular bucket in which we were riding, a yellow steel-mesh, open-air box that someone had the audacity to name a gondola. When moved over the forest, we were flying at what seemed like great speed in a big sweeping arc above the tip-tops of a host of tree species: Douglas firs, western red cedars, western hemlocks, silver firs, western white pines, and more.

Andrew Baker, a strapping Alaskan with an easy grin, told me that the fantastic sensation of speed was mostly an illusion. "I figured out the length of the arc and how long it takes us to get to the other side. I think we're really only moving at about fifteen miles an hour." Baker, if anyone, should know. This canopy crane was Jerry Franklin's idea, and a scientist named David Shaw supervises the project. But Baker, who once had been training to become a Forest Service silvaculturist, changed careers to become the technician who deals with the day-to-day workings of the business end of this crane. He serves as its safety officer (he's fully prepared, he says, to lower us, rope-wise, to the ground and then rappel down himself, should the electric power go out) and generally assists the scientists who use the crane. His official job title may be the most delicious in science and technology. He is the staff arbornaut.

The crane exists because scientists like Franklin and Shaw have become convinced that it is impossible to study a forest by looking only at what is happening at ground level, comparing it to the problem of a doctor attempting to understand a patient by looking only at his feet and legs. It is in the canopy, after all, where the tree grows upward and outward, where it captures sunlight and transmogrifies it into sugars via photosynthesis. It is where budding and branching occur, and where trees—what Franklin Delano Roosevelt once called the lungs of the planet—exchange gases with the atmosphere. By the time of my visit early in 1998, scientists were using the giant crane for some forty-five research projects, ranging from surveys of insect abundance and their patterns of distribution, to studies of how the canopy chemically and physically alters rainwater before it reaches the forest floor, to studies linking the degree of air turbulence in the canopy to the dispersal of harmful bark beetles.

As a research platform, this high-technology crane is a direct descendent of some of the earliest, and most electrifying (and comparatively low-tech), research pioneered in the Andrews Forest. For that matter, much of today's canopy research worldwide, from British Columbia to the Central American tropics, springs from early work by the ecosystem team in the Andrews, for it was there that scientists first found compelling reasons to venture into the high canopy, and then solved the daunting problem of just how to accomplish it.

On this day, Forest Service wildlife biologist Kathy Flick was conducting a routine aerial survey of songbirds in the tree canopy, a job that involves listening for their songs more than catching sight of them. It is part of a study headed by biologist Dave Shaw aimed at showing how songbirds distribute themselves vertically in the old-growth canopy, an echo of Robert MacArthur's early, and landmark, ecological studies on how wood warblers adopt varying niches in the canopy. Such knowledge could be critical, Shaw has pointed out, for understanding the habitat needs of bird populations that are in decline, or might decline in the future, in woods such as these.

Back in the crane's control booth, Creighton was controlling our movement from a computer, where a mapping program offers precise coordinates. That allows researchers to repeat the surveys like this one in precisely the same spots every time. The ability to be so precise is not a trivial issue. The crane, dubbed the Wind River Canopy Crane, is so immense that its boom can encircle fully six *acres* of forest. But that's only the beginning of its reach. The gondola can stop at any point along the nearly three-hundred-foot boom, or jib, meaning it offers, for starters, access to any two-dimensional point within (or just without) that huge circle. But there's more.

Baker, our arbornaut, murmured, "Okay, Mark, we're done here," into a microphone pinned to his lapel, and we almost instantly plunged straight and smoothly down, then eased to another stop deep in the midst of the old-growth canopy. The gondola was nearly resting on a limb of western hemlock, and we were suddenly enveloped in verdure that obscured the surrounding mountains. With the wind suddenly blocked, the air was palpably warmer. The gondola, and any researchers aboard it, can, in fact, travel not just the breadth of the six-acre circle, but up and down to any point within a cylinder. Adding this third dimension, it means their access to the canopy comprises a research cylinder of some fifty-eight million cubic feet!

Jerry Franklin began promoting the installation of a crane in the early 1990s, based on a firm belief that the canopy of old-growth forests remained one of the most important, but least explored, natural systems on earth. The oldest trees here began growing, probably after a fire, just about the time that Columbus landed. In fact, the trees here are so old that they serve as the foundation, the structure, upon which an entire airborne ecosystem thrives. Their boles and limbs have existed here for so long that they serve as substrates for other forms of life. Shaw goes so far as to suggest that the old-growth canopy is an ecosystem directly comparable to one of the most important habitats in oceans, where living organisms support not just themselves, but provide the structure on and around which a host of others grow, or hide, or feed.

Says Shaw, "The medium is air rather than water. But essentially it's a system driven by space and light, just like the marine environment. To my mind, the old-growth canopy is very much like a coral reef."

Nevertheless, here at the height of winter, it was not much of a day for locating songbirds—in fact, we were able to hear or see only four species: several clusters of red crossbills, a winter wren, a varied thrush, a brown creeper. A morning in spring would yield two dozen or more species, from giant pileated woodpeckers to tiny Vaux's swifts. But I had come especially to get a close look at something that might seem more mundane—the mosslike lichens that abound here in the canopy. Especially near the top of the canopy, we'd seen a profusion of so-called pendulous lichens, long light-green masses of tendrils that hang from branches like narrow, pointy beards. Now, deep in the canopy, in a habitat that is shadier and moister, Baker pointed to the very lichen I had come to see: *Lobaria oregana*. Virtually invisible from the forest floor, *Lobaria* were growing in great clumps everywhere on boles and limbs here: folded leafy masses of living tissue growing thick, like scalloped layers of lime-green icing on a cake.

Baker had already shown me small swaths of browning needles that had been damaged by a parasite called dwarf mistletoe. It intrudes into the very core of branches and saps nutrition from the tree, reproducing and moving through the canopy via a form of rocketry: it can shoot its sticky seeds as far as fifty feet.

A reasonable first guess might have been that the green, leafy *Lobaria* were parasites, too, sapping critical nutrients from the trees. But one of the very first breakthrough bits of research in the Andrews showed that these unassuming organisms—the same sort of mutualist fungi-alga that so fascinated Eugenius Warming a century ago—have much to do with why such magnificent trees thrive here at all.

In fact, it was because of hints from the Andrews that *Lobaria* might serve a vital role in the ecosystem—hints that came in part thanks to an old miner's lamp and a plastic garbage bag—that scientists first ventured into the canopy in the first place.

It was 1969, and botanist Bill Denison was in the Andrews Forest not looking up at the canopy, but where he thought he was supposed to be looking: down at the ground. One of the charter members of the Andrews research team, Denison had begun, even before funding arrived from the International Biological Programme, to try to develop a strategy for what he had already agreed would be his role in the whole ecosystem study.

By the time I met him in the late 1990s, Denison was approaching age seventy. He had a flowing mane of gray hair, a full silvery beard to match, and a booming laugh. As an undergraduate in the late 1940s, Denison had planned to become a physicist. But after his roommate at Oberlin College in Ohio talked him into taking a class from the ecologist Paul Sears (of *Deserts on the March* fame), Denison got hooked on the study of life.

"It was against my better judgment," he says. "I used to walk by the botany building and see those students hanging around the stairs and say to myself, 'You're never going to catch me in some musty old attic studying hay.'"

In fact it was worse than that, for Denison would spend much of his professional life fascinated by lichens: simple algae and simple fungi. By 1969, he was fortyish, rail-thin, with hair that flowed to his shoulders and a penchant for stirring things up at a university that he says was then a bastion for conservatism. He had come to the University of Oregon in 1966, after twelve years of teaching biology at Swarthmore.

"Imagine finding yourself," he says, "in the middle of the Vietnam War, moving from Quaker Swarthmore to a place that was the West Point of the West." Denison promptly helped set up an off-campus military draft resistance center (and thereby, he says, earned the lasting enmity of the conservative chairman of his botany department).

But Denison soon met Forest Service ecologist Jerry Franklin, and he quickly volunteered to become part of the core team that proposed to the International Biological Programme the plan to develop a model for how an old-growth forest ecosystem worked.

"I was assigned to the dirty-rotters," he says, with the booming laugh. His initial role, simply put, was to establish how dead organisms and detritus broke down in the old forest, and especially how the critical nutrient nitrogen cycled through the system.

As he surveyed the forest, he quickly realized that when it came to nitrogen, he confronted a major problem: there was no obvious source that could provide the critical nutrient in a form that living plants and animals can use.

"Any gardener knows," says Denison, "that nitrogen is something you've got to have if you want healthy plants." After carbon, he says, and, of course, the hydrogen-oxygen complex that makes up water, "nitrogen is the material of greatest need if you're going to try growing anything."

Living things in the ecosystem need nitrogen to form structural proteins and vital enzymes. Fortunately, nitrogen is abundant on Earth, nearly to the point of being ubiquitous, and comprises about 78 percent of the atmosphere. Yet few living things can use it directly and instead must encounter it in the form of nitrites or nitrates. Thus, plants rely on specialized organisms in ecosystems to "fix" nitrogen—that is, convert it to ammonia and other compounds, which in turn can be oxidized into nitrites and nitrates by specialized chemosynthetic bacteria in soils. Typically, bacteria that live within nodules formed by certain kinds of plants, including legumes, are responsible for fixing nitrogen from the air. In the forests of the Northwest, red alder trees play such a role: nodules on the roots are jammed with nitrogen-fixing bacteria called *Rhizobium japonicum*.

But alders normally appear only in bright sun in such areas as stream banks, or as an early successional stage in the recovery of a burned or otherwise disturbed woods. The Andrews old growth held virtually no red alder. Nor did there seem to be any other prominent nitrogen fixers. Bacteria in soils do provide some ammonia by breaking down amino acids in detritus and dead plants and animals. But by Denison's calculations, there simply didn't seem to be enough potential nitrogen from that source to sustain a forest for centuries.

"We were left with a really big question: where was all the nitrogen coming from?" says Denison. A quick chemical analysis showed, surely enough, that there was abundant nitrogen available in the soil. But there was little evidence that nitrogen was being fixed anywhere in the system.

Denison says that he found himself collecting from the forest floor loose bits and scraps, and sometimes large sheets, of a foliose lichen. The lichen, which he soon identified as *Lobaria oregana*, clearly had fallen from the old-growth canopy.

"The stuff was everywhere," as he puts it. And still perplexed by the lack of nitrogen fixers in the system, he began to wonder if somehow the ecosystem's nitrogen fixers were not in the soil after all, but in the sky. Was it possible that a lichen growing in the canopy could be playing the role of nitrogen fixer, and then raining the nutrient down to the soil?

By chance, Denison found himself with an easy way to check. Harold Evans, a prominent botanist, the university's first scientist elected to the National Academy of Sciences, occupied the office next to Denison's. Evans had recently pioneered a new technique for identifying the enzymes linked to nitrogen fixation. If exposed in an enclosed chamber to the gas acetylene, those enzymes will convert the gas to ethylene.

Denison applied Evans's technique. He had no need, however, to acquire expensive equipment to run such tests, for Denison was a skilled gadgeteer. He still owned an old acetylene miner's lamp he had acquired when he had explored caves as advisor to Swarthmore's "outdoor club." (Once widely used as light sources for everything from automobile headlamps to dangerously flaming lights in coal mines, such lamps drip water onto carbide, which react to produce the bright-burning gas.) He promptly rigged the lamp to produce a stream of acetylene and injected it into his impromptu "chamber," a clear plastic bag filled with *Lobaria* tissue. When Denison carefully vacuumed the gas from the bag and analyzed it, he found that it had converted to ethylene. *Lobaria* indeed appeared to be a powerful nitrogen fixer.

Armed with that evidence, Denison spent the spring of 1970 continually looking up, wondering how he could find out more about what was going on with the symbiotic organisms in the unexplored canopy: How much *Lobaria* was up there? How and where did it grow? And, significantly, was there some link between the age of a tree, or its forest, and the abundance of the lichen? One thing was certain: Denison was not going to be able to answer any of those questions without looking in detail at the canopy itself.

Before then, scientists had attempted to study how trees branch and other canopy phenomena by examining younger, smaller trees: canopies that can be reached from a large stepladder, or a brief climb into the boughs, at worst. But simply because he was only finding the scraps of the lichen in stands of old growth, Denison suspected that *Lobaria* might not even begin prospering until a tree canopy was hundreds of years old. As a consequence, it meant that any attempt to study *Lobaria* would somehow involve getting access to a mature canopy where the first branch did not occur until 150 feet off the ground. (As a point of reference, a reasonably mature, "large" elm on an urban street might rise forty feet to its very top. Denison was faced with going nearly four times that far simply to enter the *bottom* of the old-growth canopy.)

Still, there appeared to be an answer. If he could not study the canopy in the air, he *could* have it brought down to the ground. That summer of 1970, he says, he convinced the local U.S. Forest Service ranger to assign him an experienced "tree faller"—an expert lumberjack—expecting he could study in detail a canopy that had been freshly brought to earth. So, one summer morning, Denison found himself standing with a team of assistants safely hundreds of feet away from an old Douglas fir that was in the process of crashing to earth.

"It was like somebody had dropped a locomotive from one hundred feet. The boom when it hit the ground was so loud that I felt it more than I heard it. And the thing just exploded. We ran up and tried to examine the canopy, but there was really no hope at all. There were branches and lichen and moss scattered everywhere, and no way to tell what had been where."

At that point, actual access to study and measure the airborne canopy organisms "looked an insuperable problem. It looked to just about everyone that it was going to be impossible."

But a breakthrough came in the summer when an undergraduate named Diane Tracy, who was working for him as a summer research assistant, came into his office, made halting small talk for several minutes, and then hesitantly announced she was an experienced rock climber. She suggested that she and Diane Nielson, a fellow student worker with rock-climbing experience, could get safely aloft using the techniques that got climbers up impossible, vertical cliff faces. Within days, they had proved it in the Andrews, pounding steel bolts into the side of a giant old-growth fir and using webbing and clamps to hoist themselves steadily up.

The two students convinced Denison that the process would work, and convinced him that it could be done safely. But as soon as he became excited about the possibilities new canopy access would offer, he realized that conservative Forest Service administrators, who ultimately had to approve any new research approach in the Andrews, would be resistant to such a startling notion. So, he says with a grin, "I played to their sense of machismo," contriving to allow his students to climb a forest giant that first time without formal bureaucratic approval. "I knew that once they had done it, the good old boys wouldn't be able to admit that something two women had accomplished was too much for anyone else."

Denison and his assistants got the approvals they needed. Suddenly, and for the first time, the high canopy was an open frontier to science. The surveys that resulted from that first series of successful ascents showed that indeed the boughs and limbs of the old firs and cedars were draped with staggering amounts of *Lobaria oregana*.

George Carroll, a University of Oregon mycologist (fungus specialist), who himself had once been a student of Denison's as well as his undergraduate research assistant at Swarthmore, soon found himself involved in the early studies. "Once you got up in the trees, it was evident there was just tons of this stuff. It was all over up there."

But now Denison and his climbing team faced a new problem. To understand just how important *Lobaria* might be to the forest's nitrogen cycle, they needed to determine how much *Lobaria* might be in an old-growth canopy. An ecologist attempting to answer a question like that in, say, a prairie ecosystem would have a relatively easy time of it. Although surveying every square inch of a vast area would be impossibly time-consuming, a researcher could find an accurate answer by simply sampling random plots throughout the area, then using statistical models to extrapolate to the entire system. But doing that relies on the fact that such a prairie itself had already been surveyed and carefully characterized. (For instance, if the researcher found a certain species most abundant in low, moist swales, he would need to know how many such swales existed on the landscape.)

No one had ever done such baseline work on an old-growth canopy. If it had been done, Denison and his team could have simply surveyed parcels of the canopy and extrapolated from there. But since there was no baseline model, Denison realized that he and his colleagues would have to develop one, and not in the two-dimensional space of the prairie, but in the three-dimensional space of the canopy—some of it occupied by the tree, some of it by nothing but air.

Denison learned to ascend ropes into the canopy himself (he made his most recent ascent to celebrate his sixty-fifth birthday), but he largely relied on younger and more agile research assistants. For months, the team collected detailed data from hundreds of trees, from thousands of branches and needles, enough to extrapolate to the whole tree the nature of canopy structure, right down to how many individual, tiny needles it would bear. (An old-growth Douglas fir about two hundred feet tall has some sixty-five million individual photosynthesizing needles, he says.)

After weighing and measuring pound after pound of *Lobaria*, checking it with acetylene, and tabulating their statistical data, the researchers were able to prove that the unassuming lichens serve as critical biochemical refineries that process nitrogen from the atmosphere for the entire ecosystem.

In fact, Denison and his fellow researchers have calculated that, when all of its water is extracted to provide an accurate measure of its true biomass, nearly a quarter ton of *Lobaria* prospers in the canopy of some single acres of old-growth forest.

"I don't think there's any doubt," Denison says, "that *Lobaria* is a major contributor of nitrogen to the forest. In fact, I think it's the largest single system [for nitrogen fixation] that's been identified in the old-growth forest." He estimates that in stands of forest where it is extremely abundant, *Lobaria* can fix up to twenty-two pounds of nitrogen per acre of old-growth forest each year. The forest needs roughly five pounds per acre year to maintain itself.

"Interestingly, though, much of the nitrogen doesn't even get out of the canopy—it's taken up by the grazing of canopy organisms, bacteria, and fungivorous [fungus-eating] critters," he says.

The kicker, he adds, is that, just as he had suspected, *Lobaria* do not thrive in young forests. They appear only in stands starting at about one hundred years of age and do not abound until a woods reaches old-growth status, perhaps another hundred years.

Taking a broader look at succession and nutrients in the forest, Denison says the story that has begun to emerge is one of how such a forest sustains itself over time. If a major disturbance like a furious tree-felling windstorm, or a sweeping fire, clears out a patch of old-growth woods, the regeneration of the open, sunlit clearing would begin in a completely natural system with bursts of growth of legumelike, nitrogen-fixing shrubs, called snowbrush, then pioneering, and also nitrogen-fixing, red alder trees. "By the time these alders peter out at thirty to forty years old, you've banked enough nitrogen into the soil to take care of the system for awhile." Awhile, in this case, is a few centuries—until the forest is old enough that the *Lobaria* can return.

Conventional logging practices in the region often begin with using controlled burning to suppress the growth of brush after a forest is logged, in order to reduce competition with the newly planted tree seedlings. The new young forest that grows from the planted seedlings are typically logged again in perhaps sixty or eighty years—long before they reach old-growth status. That combination, in Denison's view, could prove a long, slow, silent nutrient disaster. Removing the early-colonizing but nitrogen-fixing shrubs and alders could reduce initial nitrogen put into the soil. Cutting trees before they are old enough to harbor *Lobaria* means that the nitrogen-fixing lichens will never drape the boughs.

For one or two or even more rotations of logging and replanting, the nitrogen already banked in the soil may be enough, he says, to sustain the woods. But in the long run, he adds, continual rotations may steadily, silently deplete the soil of nutrients the trees need for growth.

It all means, says Denison, that "if we don't care about our grand-kids, we should continue doing what we're doing. If we do care, what we're doing may be managerial folly."

The *Lobaria* discovery led George Carroll down a strikingly differ-ent research path. It began, however, with the same question. Carroll reports that as the *Lobaria* story unfolded, he began to wonder if there were other nitrogen fixers high in the air. As a fungus specialist, he wondered if there might be tiny, even microscopic, nitrogen-fixing fungi living on the surface of the tree's tiny needles. But in the process of completing a first step toward looking for those fungi on the outside of the tiny leaves, Carroll made a startling, unrelated dis-covery on the *inside* of the needles.

He was merely trying to be a careful scientist: to make sure that any nutrients he analyzed were in fact coming from fungi growing on the surface of foliage. In order to be sure that his data were not skewed, Carroll set out to be sure that no fungi happened to be find-ing their way inside the needle. So Carroll sterilized the outsides of a batch of needles to eradicate any surface-dwelling fungi. He then sliced them up and cultured them in a rich medium of V-8 juice, Wheatena, and oatmeal. He knew that any interior fungi would multiply and become apparent in such a medium.

When he returned to his culture dishes, he was astonished to dis-cover that fungi not only were present, but abounded. He tried other samples and got the same results. Clearly, not only a few, but many fungi were living *inside* the needles as a routine matter. He had, in fact, discovered a new species, its very existence—much less its role in the ecosystem—unknown to science. He named the organism *Rhabdocline parkeri*, in honor of a Canadian mycologist A. K. Parker, who had specialized in the genus.

Initially Carroll wondered if these tiny endophytes ("within plant," from the Greek) somehow also were fixers of nitrogen, per-haps processing the nitrogen contained in rainwater that fell on the needle, then somehow sharing it with the tree or the ecosystem at large. He quickly decided that it was improbable: there seemed no obvious way for an organism inside the needle's cuticle to process chemicals in rainwater effectively on the outside of the needle.

Obviously, the fungi were infecting the interior cells of these tiny, waxy needles for a reason. But why? From the perspective of the fungi's need for survival, the reason itself seemed obvious enough. These were not lichens. They had no algal partner that could photosynthesize, packaging energy from the sun as sugary photosynthate. Since fungi can't photosynthesize, they were probably "stealing" photosynthate from the needles—eating the very sugary food that the needles produced. This raised a curious question: If fungi were parasitizing these needles, why didn't the needles show any signs of wilting or, indeed, of any evident damage at all, as do the needles of trees infested with dwarf misletoe? Why, for that matter, did the trees themselves seem, paradoxically, to be thriving?

If the fungi, by parasitizing the needle, weren't actively harming it, perhaps they were only extracting a minuscule bit of food. Perhaps they had evolved not to expand their population, even in the presence of abundant food, in order to kill the host that fed them. But Carroll wondered if there was a possibility that the fungi were somehow doing the needle some *good*.

Based on the behavior of endophytes in more familiar plants, like grasses, Carroll hypothesized that the endophytes he found are a sort of chemical-weapons factory used by the trees in their ongoing evolutionary war with defoliating insects. Indeed, he quickly came to believe that the fungi provide an absolutely vital—even life-saving— role for the trees. It appears that these fungi, like the mycorrhizal fungi of the roots, are mutualists—symbionts. In exchange for the energy-rich sugars and starches fed to them by the tree, they return the favor by forming alkaloid compounds—poisons that act against defoliating insects, a sort of chemical warfare conducted on behalf of their host. Short-lived, like the insects, the endophytic fungi biologically adapt to new attacks quickly, evolving new defenses even as insects evolve new tolerances against the old chemical warfare.

Many plants produce poisons to ward off pests. For instance, caffeine in the coffee tree, nicotine in the tobacco plant, or functionally similar chemicals like digitoxigenin, a steroid produced by the foxglove plant, all serve to ward off insects (and often turn out to have medicinal effects on humans: in the case of the foxglove alkaloid, it is used as digitalis to shock-stimulate life back into heart attack victims. Caffeine attacks the anxiety center of the brain—just enough, in the minuscule doses typical of a cup of French roast, to enliven the senses and energize the morning).

A small, annually seeding plant that might live only months or years before reproducing can evolve changes in its chemical defenses, a critical ability considering how quickly short-lived insects can evolve and adapt defenses against the defenses. But trees—and especially trees like Douglas firs that have extremely long life spans—face a unique problem. They simply live too long to adapt in synchrony with short-lived insects. And yet they must have a strategy to survive.

Consider the problem they face. Working with bacteria—in this case a common microorganism *Escherichia coli*—researchers have been able to demonstrate rapid evolutionary adaptation to attacks by toxins. Cultured in a simple petri dish, a single bacterium can become a pile of several million in a single day. Dosed with an antibiotic, the colony virtually vanishes in another day. And yet, in typical experiments, a scattering of resistant bacteria tends to survive—perhaps only one or two or three. But in another day, these few genetically resistant organisms reproduce into a new heap of millions of organisms. And the population will need a much higher dose of the same antibiotic, or a different antibiotic, to suppress it. (This very phenomenon has many medical biologists deeply worried about the rise of increasingly resistant organisms, including the microbes that cause such dread diseases as tuberculosis.)

Similar experiments have been run repeatedly with insecticides in farm fields and homes and businesses. Even common house flies today carry genes that make them resistant to doses of such insecticides as DDT, which would have easily wiped out almost all of their kind a few decades ago.

"Since insects can adapt to any defense so quickly, ecologists like to ask the question: why are there any trees left *at all* in the world?" says Carroll. And certainly, he says, how old-growth trees can live so long—the very trees that provided the superstructure for the growth of *Lobaria*—had long remained a mystery.

The answer, Carroll realized, could be that the long-lived trees don't need to adapt because their endophytes, which are short-lived, *can*! "With these fungi inhabiting trees, you can suddenly allow the possibility of a changing chemistry," he says.

His follow-up studies supported that notion. Carroll has compared the survival of tiny, foliage-attacking flies called gall midges that had infested both endophyte-infected and -uninfected needles, and found significantly higher rates of mortality of the midge larvae in the fungus-rich environment. The fungus had not invaded the bodies of the tiny flies, so in all probability it had produced a poison that killed them. Other studies later showed that *Rhabdocline parkeri* produces chemicals that suppressed growth in or killed another pest, the spruce budworm.

Studies of endophytes living symbiotically with other plants support the notion as well. Some grasses, notably the fescues, can support endophytes that can have startling effects. These fungi produce vasoconstrictor compounds—that is, chemicals that can constrict blood vessels. In some cases, cattle grazing on endophyte-infested fescues have quite literally had their hooves and tails fall off for lack of blood supply. (The pharmaceutical industry uses similar vasoconstrictors for medicinal purposes, for the control of migraine headaches, for instance.) The vasoconstricting toxins not only have a profoundly discouraging effect on hungry insects, they also can cut off blood flow to the extremities of grazing mammals—to the point that the animals can quite suddenly develop gangrene—hence the body parts falling off.

The pattern of life for the tiny fungi in the needles is patience defined. The fungi release their spores only when the needles die. The spores find their way from dead needles lodged somewhere in the canopy to a fresh green needle. The spore develops an infinitesimal spear that lances its way into the needle, penetrating and infecting a single cell. For as long as eight years (more often about five) the endophyte simply waits. It grows virtually not at all. After all, if the tiny mote of fungus were not to remain in this state of near dormancy, it would have to seize more photosynthate from its thriving needle host in order to grow.

It matures and releases its own spores only when the time comes for the needle itself to die, beginning this odd, slow cycle again. The quiet cycle continues until a needle happens to be attacked by insects. In this case, the fungi simply poison the insect and then begin to prosper, tapping the insect, rather than the needle, for nutrients.

How might this remarkable mutualism have developed in the course of evolutionary time? In the process of sexual reproduction, genes combine and recombine in new patterns and combinations. Sometimes, especially through mutation, they will produce an entirely new trait. That evolutionary process forms the basis of how species adapt, and change, and in fact sometimes speciate into entirely new forms. It is biological trial and error. The genetic dance in the cells of the mammal antecedents of the giraffe at one point somehow randomly came up with a longer neck; then, when that longer-necked offspring survived by dint of its ability to browse at an altitude no other mammals could reach, a random mutation produced a longer neck still.

In the competition between predator and prey—or herbivore and plant—each adapts in response to the other. In time, a species of plant, by recombining its genes, might develop a toxic defense against insect attack (the nectar of milkweed, for instance, is so toxic as to kill most insects). The insect, in turn, recombines genes in order to cope with the noxious chemical. In the case of milkweed, the monarch butterfly is adapted superbly to consume its nectar with no toxic effect: the monarch is the prime pollinator of milkweed, in return. Beyond that, the presence of the milkweed toxin in the monarch's system makes the butterfly poisonous to its own potential predators. The viceroy butterfly took a different evolutionary tack: although it carries no such toxins, it evolved a pattern of body coloration that mimics the "stay away from me" black-and-orange pattern of the monarch. And the dance goes on.

The fungus's short life span thus provides the toxicity or noxiousness that keeps insects from evolving, over time, perfect immunity to a particular tree's defenses and defoliating it. As the insect adapts, the endophyte adapts in response. In fact, says Carroll, a single tree's needles may harbor endophytes with a range of different insecticides or repellents in them, an overwhelming protection against a wide range of potential defoliators. Meanwhile, the tree thrives.

Oddly, after the flurry of initial discoveries, support from universities and major funding sources, like the National Science Foundation, for new research in the high canopy virtually dried up. At least part of the reason was because access was so difficult and the process of discovery relatively slow. But by 1973, Denison had published reports in both a Forest Service publication and in *Scientific American* detailing his team's progress in reaching the canopy with cliff-climbing gear. (Denison insisted that since his key abornauts were women, all the third-person pronouns in the Forest Service article be feminine, a first for the male-dominated agency.) The article, in turn, spawned a scattering of new research efforts in the high canopy of tropical forests. In the 1980s, the Smithsonian Institution installed the first canopy crane in a rain forest in Panama. French researchers later developed another technique for studying the canopy of the tropical forest in Africa, using a balloon that could drag the scientists over the top of the canopy in a basket they called a sled.

Still, most access has involved climbing. By the mid-1970s, Denison's group had refined its technique. Rather than hammering bolts into a tree for initial access, they found that they could shoot thin but sturdy monofilament fishing line over a low branch with a powerful slingshot or, even better, a crossbow, spinning the line out of a rig holding an actual fishing reel, then using that line to haul up a heavier nylon cord, and then a strong climbing rope.

On a humid autumn day in 1995, I sampled the technique in the Andrews. It was enough of an experience to prove the value of an elevator—or its canopy analogue, the skyscraper crane. Nathan Poage, a limber and enthusiastic young master's degree candidate, had offered to show me the approach.

We hiked upmountain, through soaring old Douglas firs, hemlocks, and western red cedar. We stepped in the humid autumn air along a moderately worn path, through bunchberry and rhododendron, Oregon grape, salmonberry, sword fern, and a nasty plant they call the devil's club, *Oplopanax horridum*, with huge, piercing spikes on its stem. Whitewater roared over boulders and logs in Lookout Creek below us, and we climbed steadily, moving upstream, at one point encountering and then neatly stepping up a stairway some enterprising soul had hacked into a centuries-old giant log that lay across the path.

Poage, a graduate student in forestry at Oregon State, found the tree he was looking for at a spot where we were up to our hips in huckleberry. Beside the tree was an ancient snag, a bit of giant tree that had snapped off maybe forty feet from the ground. Like a sort of headdress, a broad, bushy green fern burgeoned from its busted top.

Our tree was an even greater giant. Sweating, we dropped our backpacks to the ground. While Poage pulled what seemed like miles of coiled rope from the packs, I made measure of our tree. Carefully I moved three times around with my six-foot span of outstretched arms. My fingers did not reach my starting place. Thus, by my estimate, the tree was some twenty feet in circumference, some six feet in diameter. Hollowed out, the bole would provide a circular room with about the square footage of a small office, twenty-five stories high.

The tree was already rigged for climbing. A thin yellow cord, like a piece of light nylon clothesline, ran from here on the ground to the tree's canopy high above, then all the way back down to the ground in a loop. Poage rigged the end of an immense pile of blue rope from his backpack to one end of this smaller white rigging line and then began pulling the other end, using the weight of his body increasingly against the thick blue rope's increasing weight as he hoisted, hoisted—some 150 feet of rope up to a branch and then another 150 feet back down, the hoisting easier now. The blue rope, although heavier than the thin yellow line that had hauled and guided it up, was the thickness of only a finger.

In another moment, Poage had climbed into a special harness that had him hanging, his rear end just inches from the ground, in a sitting position. He was bobbing on the blue rope like a human yo-yo, for it was a special elasticized, shock-absorbing, "dynamic" climbing rope that functioned not unlike a stiff, immense rubber band. Two loops hung from the harness he was wearing, one for each foot. In a couple of motions Poage described as dolphinlike, he swam up the bouncing blue line, standing first in one loop while he slid a special clamp called a Jumar, which was attached to the loop around the opposite foot, then alternately standing, sliding up the opposite clamp and loop arrangement, sitting. Then sitting quickly again, standing in the opposite loop, sliding the other Jumar upward, and quickly repeating the action, he was suddenly ten feet above my head, still gently bouncing. And then he was off, dolphining his way up the rope thirty stories up, pausing for a brief rest; then sixty feet up, then suddenly ninety. Minutes later he was a tiny figure against the canopy's ceiling of green. Poage worked himself to the first branch and then, with a quick upward jolt, vanished into the verdure.

A few months after my visit, a special platform was installed in that very giant of a tree. Designed to last for the long term, its installation involved no nails. It was virtually strapped to the tree, the straps loosened each spring to accommodate the tree's own bit of annual radial growth. The idea was to provide a place for setting up monitoring instruments and for researchers to work from. The platform, which was modeled after a similar platform system set up by researchers in the wet old-growth forests of British Columbia, is large enough that small groups of researchers have since camped on it, swaying in the Cascades wind overnight.

Climbing can take hours for the inexperienced. Although the view on the way up is magnificent, it is not, as I learned in a partial sample ascent of about forty feet, for those inclined to either vertigo or motion sickness. Botanist Art McKee, a charter member of the Andrews team and now the laboratory-forest's on-site manager, says he was determined to accomplish the climb during Denison's team's 1970s canopy research heyday.

"It took me something like two hours to get up to the canopy," he says. After the first ascent left him with vertigo and "cotton mouth," he decided to prove to himself he could do it again, without the initiate's terror. He climbed on a gusty day.

"The tree was swaying in the wind. When I got up to the canopy there was a research assistant up there, sitting on a limb and eating a sandwich. She said something like, 'Isn't it wonderful?' I couldn't answer her. I had cotton mouth again."

My own sample of the climb was not an elegant effort. I climbed high enough to peer down into the top of the broken snag, a wide, round, tilted table, where there was not only the huge fern I'd seen from the ground, but clumps of moss and small plants—a virtual garden that had been developing for years, perhaps decades, on the broad table at the top of the snag. It was not until my arms were exhausted and bruised from attempting to chin my rather hefty self up the rope, and then managed somehow to snag my Jumars—the climbing clamps—one into the other, that the agile Poage suggested that something was wrong. He had borrowed my harness from a young woman nearly a foot shorter than I, and now he realized that the loops were many inches too short for my height, making it nearly impossible for me to use my legs, rather than arms, for climbing power. Hanging beside me on an adjacent climbing rope, Poage jury-rigged a solution, making loop extensions out of bits of spare webbing. And I discovered, indeed, that it was now an easy matter to simply use my *legs* to propel myself up into a stance. But by then another unanticipated problem had developed. The rope, which hangs several feet from the trunk, bounces and sways, and the hapless climber also tends to spin, corkscrew fashion. It is a symphony of motion that makes a ride up and down sea swells seem like a walk on solid ground. A slow forty feet up, I was nearly blind with motion sickness from the combined effects, and I descended with at least a new appreciation for the rigors of rope-based research.

Nalini Nadkarni, who was a graduate student at the University of Washington in the 1980s, was the first canopy researcher to use rope and harness to climb into the canopy of broad-leaved trees in the Northwest, rather than conifers, in this case big-leaf maples in a temperate rain forest on Washington's Olympic Peninsula. To her astonishment, she found high in those branches not only what she expected to find of the tree—leaf, bud, and branch—but something she says she never guessed was even possible. She ascended merely looking for epiphytes. Epiphytes are the "air plants"—those that depend on trees or other plants for support, but not nutrients. Epiphytes include various vines and the tropical orchids.

"I wasn't prepared for what I found. I just wanted to see if I could figure out how many epiphytic materials I could find up there. But I kept cutting off bits of moss and finding these roots. So I traced them back, and found out that they came from the *tree*. So I went back [to the University of Washington] to my major professor and asked him to explain those weird roots growing out of the tree up in the canopy. And he said, 'Huh?' "

Her advisor, it turned out, had never heard of any tree, anywhere, doing such a thing. Nadkarni sent a flurry of letters to other biologists around the country. She found she had made a heretofore unknown discovery: that some large, old, broad-leaved trees can develop adventitious roots into their own canopy—into the "soil" that has built up on giant limbs from the growth and death of epiphytes, animals, and its own shed leaves and other detritus. Now a member of the faculty at Evergreen State College in Olympia, Washington, Nadkarni has discovered the same phenomenon in the Monteverde Cloud Forest of Costa Rica, where she now spends several months each year conducting canopy research.

But she says the importance of her initial discovery—remarkable enough that it soon made the cover of the prestigious journal *Science*—goes far beyond showing us something new about how trees function. It hints that there are many more startling discoveries to be made, with implications for the way both trees and forests as a whole work.

"If we're really interested in the whole forest, we really need to get up to where the leaves are," she says. "We can only get certain kinds of information if we study organisms out of their natural environment. It's like studying elephants in the wilds of Africa instead of in a zoo."

The gaps—the holes in knowledge—are presently almost unbelievably vast. Nalini Nadkarni puts it this way: "At first canopy science was viewed as almost a sort of Tarzan, semi-science, in the same way scuba divers were viewed at first by marine biologists. It's exciting work. But there hasn't always been a lot of scientific rigor in canopy science. For instance, very few studies have been duplicated, because of the access difficulties."

And there remain, she says, "tremendous holes in knowledge of the canopy. For example, we know that microbes rule the world. But we know almost nothing about microbes in the canopy. We don't even know if they're different than the microbes found in the soil."

As a research frontier the canopy has revealed itself as a wide open one. "That's where the highest level of biotic activity is," says David Shaw. "That's the interface between the biosphere and the atmosphere and the forest ecosystem. It's where the budding and branching occur. That zone of interaction in the upper forest canopy is where the highest level of photosynthesis is going on. Light is the driver of the entire ecosystem. Photosynthesis in those needles is turning light into sugars. But it is the area of the entire ecosystem— in the entire terrestrial *biosphere*—that we know the least about, even though everything else in the forest is an add-on in terms of what's happening right there, at the outer forest canopy."

For the Andrews team in its early years, research in the high canopy was hinting at an even more fundamental issue. Through simple reasoning, Denison and his team had hypothesized that the canopy held unknown ecological treasures; through difficult and detailed work, they had proven their hypothesis true.

It was the first clue that perhaps these ancient woods hid other surprises. In fact, it foreshadowed a cascade of discoveries, secrets of these old woods that have everything to do with the mystery of how the forest functions, thrives, survives.

5

THERE ARE SOUNDS, all the time, in the forest. On one autumn day two years after my first trip to the Andrews, I returned near sunset to Reference Stand Number Two, the site where Fred Swanson and Lynn Burditt and I had stood on the log. I climbed the same fallen log, stood there, and simply listened.

It was a quiet evening. The boughs overhead barely stirred in the breeze. But if I could have listened with just a bit more acuity, I would have known for certain that the forest was far from silent. Over time, it resounds with matter falling from the trees. The forest clonks and bangs and sings with the hissing and the booming and the knocking and the thwacking of bits of litter, from pieces of limb or lichen, moss or needle, of seed cone or sloughing bark. For hour by hour, day by day, the standing forest ecosystem fairly rains down bits of its own life, and death, to the forest floor.

Ecologist Elliott Norse has called the fall of litter "the capillary system of the forest, removing wastes and conveying food to the legions of consumers below." The sheer variety of stuff that falls to the forest floor is extraordinary, from bits of caterpillar frass to the virtually invisible spores of fungi, from shreds of mammal dander to the occasional bird feather, from the droppings of flying squirrel to a gossamer shred of spiderweb to bits of the waxy exoskeleton of a mite rejected by the spider that just ate the rest of the tiny arthropod. In the short time that one stands on a log and listens, it might not seem to amount to much. But it is a magnificent capillary system: in a year, onto a single acre of old-growth forest, some five tons of litter will fall.

The rains and the snows, too, and even the drip from fog, provide major transit for nutrients to flow from the canopy to the forest floor. The precipitation dissolves nutrients in mosses and fungi. Bacteria, for example, that live as "lawns" on the exterior of *Lobaria* wash off and fall to the soil as nitrogen-packed parcels. The water drains and drips those nutrients back down to the soil. (Although not always: it also can serve to transport nutrients from one place in the canopy ecosystem to another, as it dribbles and runs down branch and bough.)

The litter, and precipitation-borne matter, links the ecosystem of the canopy to the ecosystem of the forest floor. Once it reaches the floor of the forest, the litter will become fodder for the creatures of decomposition and the organisms of decay, from munching strong-jawed shredders to nutrient-dissolving fungi. And eventually, these chompers and munchers and decomposers will release the very nutrients back into tree roots that allow the tree, and the canopy from whence the litter itself came, to prosper. The tree, in essence, helps to feed itself. More accurately, the tree is a virtual ecosystem in its own right, cycling its own components through life and death, decay, and decomposition, and back into life again.

These billions of bits and pieces are hardly the only matter that fall to the forest floor. The Andrews scientists use the inelegant name *coarse woody debris* to denote litter of quite another kind. They mean big pieces of wood, like the chest-high log I was standing on, a log long enough to serve as a platform for the entire membership of the U.S. Senate.

On an earlier day, a few hundred miles to the north, walking a trail through a piece of the Hoh River rain forest in Washington's Olympic National Park, I had overheard a man muttering to his wife about the lassitude and incompetence—as he perceived it—of the National Park Service. He couldn't believe that here, in a national park, the authorities could allow such chaos. A parcel of woods that I saw as a spectacular remnant of lowland, coastal temperate rain forest, he saw as a disaster. He wondered, out loud, why Park Service employees didn't "get in here and clean up this mess."

The mess in question was the litter of giant logs that lay about, covered with moss and nursing into life young trees that grew in a row on them. The man spoke admiringly of a roadside park they'd stopped in somewhere along an interstate highway, where any trace of such a "mess" of old fallen trees and limbs had been cleaned up, where only green grass grew between giant trees: a proper park.

This tourist's attitude, in turns out, is a nearly perfect reflection of a long-held precept of the forestry profession itself, one that perhaps more than any other the whole-ecosystem study by the Andrews team came to refute: the notion that only by completely clearing a site could forestry be optimal and efficient.

For Douglas firs in the Pacific Northwest, the presumed time to log a forest and start a new rotation—that is, the time when foresters assume that the point of diminishing returns in wood production (mean annual increment) has arrived—comes at perhaps sixty to one hundred years. By time the Andrews study began in 1970, conventional wisdom held that a forester looking to provide maximum wood fiber at lowest cost over time would not only design a logging and replanting program based on this assumption, but order a logger to clean up any logged site thoroughly. Some wood unsuitable for lumber could be hauled away and turned into paper pulp. Perhaps some could be ground into chips or pulverized into sawdust that could in turn be glued and pressed into new substitutes for plywood, like particleboard and strand-board. Typically, foresters went even further and mandated that any woody litter that could not be removed and used instead be burned at the site. The debris, after all, could harbor insects and fungi that could harm replanted trees in a new, farm-style tree plantation, or might present a fire hazard to the replanted woods.

The ecological history of most forests is the history of cycles of what most of us might see as catastrophes: fires, floods, insect out-breaks, or the kind of raging windstorm that had brought crashing to earth the trees at Reference Stand Number Two. It was certainly no news to conventional foresters that disturbances occur in nature. The documented history of periodic fires and windstorms in this region have long formed much of the rationale for clear-cut forestry. In fact, a core assumption of forestry was that a natural forest could be *improved* by making a disturbance far more thorough than nature ever could. If clear-cutting was merely an imitation of a sweeping fire, what could make more sense than cleaning up the woods—and, for that matter, any streams that laced through the woods—salvaging the wood fiber in any downed logs that hadn't yet overly succumbed to rot?

In fact, that sensibility had become entrenched conventional wisdom in the forests of the Northwest by the 1960s. Forest ecologist Mark Harmon suggests part of that was simply a swing of the pendulum away from another source of excess. During World War II, he says, loggers, although rapidly clear-cutting vast expanses of forest, actually removed only the high-value, rot-free large logs. Later calculations showed that enough wood fiber was left behind in the Northwest alone to feed the entire U.S. pulp and paper industry. In the postwar years, he says, the response to that perceived waste was an even more intense, clean-it-up approach—he calls it "anal-retentive forestry."

It might seem obvious that any huge, decaying dead organism, and certainly a giant log, would be valuable to an ecosystem simply for the nutrients its decay would release. But most of a tree, and certainly most of its central bole—its log—is wood. And wood is mostly dead matter. Although rich in carbon, a log contains only a fraction of the other nutrients contained in a tree. Those reside in leaves, in buds, at growing tips of stems, and in the thin inner bark that provides transport for photosynthate throughout the organism into the roots. That fact certainly lent weight to the notion that a rotting log on the forest floor was nothing more than waste.

In few places was clean-it-up forestry as thorough as here in the Douglas fir region. Because the Douglas fir is intolerant of heavy shade, and grows vigorously in a flood of light, the clear-cut— thorough and industrially efficient—had become the logging method of choice, with the usual controlled burn and herbicide treatment. The subsequent reconstructed forest was nearly as clear of clutter as my tourist's admirable roadside park.

But clutter-free was far from what the researchers found in the Andrews. In one early survey, they weighed 219 tons of downed logs on a single acre of forest, covering one fourth of the surface area of the forest floor. There were another 47 tons of standing snags. And as it worked its way toward understanding the old-growth ecosystem, the Andrews team began to suspect that the clean-it-up model might be far from ideal for recovery in many disturbed woodlands, whether disturbed by wildfire or by the lumberman's chain saw.

"The Forest Service used to spend literally hundreds of dollars per acre to clean up logs," says Jerry Franklin. "The primary view of downed logs was that they were a fire hazard and a waste and an impediment to travel—in short, that they didn't do anything but cause problems in the forest." That opinion was even stronger, he says, if the logs had begun to rot. "Virtually everybody believed rotting logs were trash—wood fiber that had gone to waste. In retrospect, it's incredible that we could have been that stupid."

To understand the tree in its death is to first understand the tree in life. In life, a tree is not only a triumph of biology, but a triumph of nature's engineering. Cup, in your hand, a leaf. If that leaf could be somehow stretched to the size of a football field, and we could peel back just enough of its waxy cuticle, its skin, like the roof of a giant tent, and climb inside, we would discover the astonishing place where this energy is intercepted, and then stored, in the tiny, bio-chemical power plant called sugar. If we wandered around in this acre-sized leaf, we would clamber around green pillars that hang, like sausage-shaped stalactites, from the cuticle roof of the leaf and, on the cuticle floor, over sodden, spongelike cells, virtually steaming water vapor. We'd find surprisingly abundant air space between these pillars, called palisades, and around the sodden, steaming spongy clusters, called parenchyma.

It is here that the wonder of photosynthesis proceeds. It begins in specialized structures in the palisades, in green, saucerlike chloroplasts that are stacked like coins in a tube. A chloroplast is a biochemical catcher's mitt: grabbing from a stream of radiation from the sun infinitesimal pulses of pure energy. The job that these structures must accomplish is the keystone in the arch of the miracle of life. It is the feat first invented by the most primitive, single-celled eukaryotic algae in the seas during the Precambrian, some six hundred million years ago: to bundle up pure and ephemeral solar radiation into a neat biochemical package, a molecule of food that can store energy for hours, days, weeks, years. It is energy that will flow through the tree, and even beyond, into the food webs of the great forest ecosystem.

The palisade cells must package energy not only for the tree, but for the mite chewing on tiny bits of the stem, the predatory spider that eats the mite, the flycatcher that eats the spider, indeed for the fungi pulling sugar out of the tree's roots on the forest floor, or the squirrel or vole or deer that eats a bit of fungus.

The act of photosynthesis hangs on a photochemical evanescence in time. Light streams past the cuticle and into the chloroplasts at 186,000 miles per second. The catcher's mitt must seize a bit of that sunlight—a feat accomplished in a photochemical instant by special pigments in the chloroplasts.

Snap a photograph with a camera, and you engage in a photochemical instant: a pattern of light moving through the lens hammers its photons at a silver-sensitized bit of film, photochemically reconfiguring the molecules into a pattern that will become an image on a negative. Similarly, a photon hurtles its way to a molecule of the pigment chlorophyll and triggers a chain reaction, destabilizing a "light trap" molecule so vigorously, and exciting its spinning electrons so wildly, that one electron from a molecule in the green light-trap literally achieves escape velocity and spins out of orbit. All in an instant of time, that wildly careering electron hurtles its way into the orbit around an atom behind it, in turn energizing and destabilizing that atom, until one of its own electrons spins out into the next atom in the chain. The pattern replicates in the next atom, and the next, and the next, as a long, nearly instantaneous skein of energy-consequence blazes through the green membrane and into a wet solution of water and proteins. There, the sheer force of energy that moved electron to electron severs the strong bonds between hydrogen and oxygen in a water molecule. The foundation act of photosynthesis has been accomplished.

The split of a water molecule leaves free atoms of oxygen, which snap furiously together in pairs and then diffuse out of the leaf's stomata, the pores that allow it to breathe O_2—the oxygen of the atmosphere. Trees and other plants manufacture the Earth's atmosphere. Virtually all of the oxygen in the air we breathe was made by green plants, photosynthesizing over hundreds of millions of years, ripping apart molecules of water for the benefit of their hydrogen, sending the residual oxygen into the skies.

In the leaf, the liberated hydrogen proceeds through a series of tightly linked reactions with enzymes in a complex process called, with intimations of mysticism, the dark phase. Just as oxygen has diffused out of the stomata, carbon dioxide has diffused in. Now the hydrogen atoms and enzymes react with that carbon dioxide.

The finished product is a molecule made of one carbon, two hydrogens, and an oxygen. You can find it abundant not only in the sap of the living tree but, for that matter, in packets on the table of any diner. It is sucrose, simple table sugar. The sweet of the sugarcane or of the beet or of maple sap is simply an amplification of the sugar-rich bath in which every plant cell lives. It is photosynthate, the child of sun, and carbon in the air, and photosynthesis.

Scientists, in their labs, have tried to replicate this wonder. To a degree, they've succeeded. With enormous, detailed effort and excellent equipment, a trained scientist can indeed produce a few motes of sugar from the same basic chemical recipe. But perhaps an ornamental maple sits outside the window of that clever scientist's laboratory.

If that tree happens to be mature, and in full leaf, it offers to the sun, in layers, about half an acre of total leaf surface. That half acre of leaf absorbs some 188 million calories of sunlight per hour, just under three *billion* calories on a long midsummer day. If each square meter of leaf is photosynthesizing optimally, it produces about a gram of sugar per hour. Over the course of one growing season, the biochemist might dither with high-tech equipment to produce an iota of sugar. In the same time, the tree silently produces about two *tons*.

A simple one-celled plant can efficiently produce photosynthate. But for a plant to be a tree it must manufacture itself into existence. It must make wood. Of all the remarkable feats a living tree performs, this may be the most astonishing of all. A tree must, like all vascular land plants, accomplish two things: It must send the water and the mineral nutrients and nitrogen it has taken from the soil up and outward to every living part. It must, at the same time, transport *in the reverse direction* its energy packages, its photosynthate, its sugar to every living part. At the same time, in order to *be* a tree, in order to compete with its neighbors, it must grow to be a giant among organisms: it must find a way to engage in a feat of natural architecture that boggles the imagination—it must find a mechanical way to support tons of its own weight, often against gale-force winds, the weight of tons of wet snow, of sodden tons of leaves. To accomplish all that, it must build—of a rock-solid foundation of main roots—its trunk, its bole (main stem), and the secondary stems that will grow out as major boughs. And it must accomplish it while spreading its leaf coverage, expanding its root system, all in well-tempered synchronization.

We know we can learn the age of a felled tree by counting the growth rings on its stump. If we look more closely, we might see that the disk is layered: bark first, a thin band of inner bark, then a thick layer of more porous sapwood, and, finally, in the center of the disk, dense heartwood.

Just inside the inner bark is the tissue-thin cambium, the organ that performs the miracle of wood manufacturing in the tree. From photosynthate and nutrients and a whiff of hormones and enzymes, this microscopic structure manufactures the plumbing systems through which the photosynthesizing leaves will sate the hunger of the living cells through the tree. This underbark structure is called the cambium, from the Latin *cambire*, "to exchange." It enwraps all the parts, from the farthest tips of roots, to trunk, boughs, stems, all the way to the highest reaching of twig tips in the canopy. On its outer surface, just below the inner bark, it will build and maintain a delicate plumbing system called the phloem, through which the dissolved sugars of photosynthesis will move inward and downward from the leaves in the canopy, down bough and bole, and out to the tips of the deepest roots.

But mother cells in the cambium will also manufacture a separate, far thicker plumbing system on the inside of the tree. It manufactures xylem, the cells that will become wood. While alive and developing, they will swell, elongate, and hollow themselves out like tiny straws. They will be composed first of softer materials—cellulose, and pectin and related polysaccharides. But then they will suffuse themselves with lignin, a tough, rigid material with qualities much like epoxy cement. Lignin will eventually comprise as much as one fourth of the volume of the wood. (It is lignin that lends the vanilla odor to fresh sawdust. Breaking down this concretelike substance to get at the wood's soft cellulose fibers is one of major challenges faced by paper pulp manufacturers.)

In the sapwood layer, through thousands of the slim, pipette-like channels, the tree will pump water from the roots, along with vital dissolved nutrients: nitrogen, phosphorus, potassium, sulfur, calcium, magnesium, and iron, and traces of chlorine, iodine, cobalt, and manganese.

Water, and its dissolved nutrient soup, must saturate the cells of living things. Even we are about 65 percent water by weight. Our blood is 83 percent water, and even our tooth enamel is 2 percent water. The salts of our blood are legacy of a long evolutionary trail that leads back to early forms of life in saline seas. Similarly, the tree is directly linked to the lives of the first in the primordial seas. The land plant exists not as an organism totally distinct from the plants of the seas but, writes botanist E. J. H. Corner, "by modification of the construction inherited from a marine ancestry." In other words, the tree must saturate its living cells in a functional, nutrient-rich sea.

The porous sapwood provides the plumbing that allows that to happen, 250 feet (in the case of an old-growth Douglas fir), even more than 350 feet into the air (in the case of a giant sequoia), far removed from the soil from which the tree must extract its water. Photosynthesis, and the physics of water, will provide the pump. Undisturbed water molecules tend to form remarkably tight chemical bonds with each other. That's the reason a razor blade will float in undisturbed, calm water in a goblet. Because of these bonds, a thin column of water, pulled up through a fine tube, has tensile strength approaching that of metal wires. But this wire needs some kind of winch. That winch can be found in the leaves. As photosynthesis rips apart water molecules, and as additional water simply diffuses through the leaves' stomata, their furious departure at the top of the wire provides a tug, pulling the water column up the network of slim pipes in the sapwood. It is a waterworks under tension. Water hauls itself up, molecule by molecule, to the outermost branches, to the tips of leaves, up to the top of the crown.

As the tree grows and ages, the vessels in the sapwood clog up and solidify. Eventually they stop carrying water altogether and become heartwood at the core.

In a living tree, almost all of the mass that is wood is dead. Yet in its death, the wood strengthens the root, or bole, or bough. For decades, or centuries, or even millennia, these same woody cells will serve as the superstructure that lifts the leaves of gargantuan plants to the sun, a frame so strong that it can compare with steel or concrete.

In short, the live tree is a meld of living, growing matter and a mostly dead, but brilliantly engineered, superstructure to support that life. As I had seen in the canopy, a tree is a veritable ecosystem in its own right, the substrate for everything from nesting birds to creeping spiders. As it grows, and branches, and turns its leaves to the sun, it is an intricate system of control and feedback that continually configures, reconfigures, and mechanically reengineers itself in response to its environment, all the while seizing photons from the sun and microbits of elements from thin air and ordinary soil to build itself as a structural giant among organisms.

And if ecologist Mark Harmon is right, it all makes the living tree not half so interesting, nor even so alive, as a dead one.

The day after my return to log-side at Reference Stand Number Two, I found Mark Harmon in a large, open shed at Andrews headquarters. It was a noisy visit, for Harmon, in a sawdust-caked T-shirt, plastic goggles, and earplugs, was in the process of conducting science with a whining, howling chain saw.

From the end of a two-foot-diameter log, Harmon sliced cookies of wood, pieces that looked like primitive wheels. He held a cookie slice up to catch the light, then, with permanent markers, began highlighting the regions of preliminary rot in the slice.

"That's blue stain," he said, pointing to an early fungal invader that bark beetles convey into a downed log as part of the first wave of centuries of attack that will eventually turn this log to dust. And he showed me white-rot and brown-rot in other samples, both actually the visible tissue of fungi.

Not far away, we had visited the site this log had come from, not randomly from the floor of the forest, but from a small clearing beside a road in the Andrews where Harmon in 1985 lined up row upon row of similar logs. Since then, every few months, Harmon has cut more cookies from the ends of the logs, marking rot, taking photographs, saving smaller chunks in plastic baggies to analyze more thoroughly in a laboratory. From these studies, Harmon is working toward determining, among other things, just how much nutrition the logs eventually contribute to the forest ecosystem, and just what role they play in the storage and release of carbon dioxide to the atmosphere. The latter issue is critical to understanding the role that forests, and especially old-growth forests, play in the continuing drama of global climate change. Carbon dioxide is a key "greenhouse" gas. Once released in the air, it tends to rise into the upper atmosphere, forming a sort of reflective barrier that helps reflect head radiation back to the earth's surface. A broad consensus has developed among scientists that a steady buildup of carbon dioxide in the atmosphere will lead to a warmer climate in the next century.

As they photosynthesize, forests remove massive amounts of carbon from the atmosphere. And although rotting logs will eventually release that carbon, they will do so exceedingly slowly, serving, in the meantime, as giant reservoirs for carbon. (On the other hand, a piece of coarse woody debris that is turned into pulp, then paper, then burned in a municipal garbage incinerator would release stored carbon rapidly.) Predicting what roles trees, living and dead, play in abating or intensifying global warming will depend on laborious, detailed studies like this one.

In short, Harmon is trying to learn how an old-growth log rots. As a research project, Harmon's effort is highly optimistic—almost hilariously so, especially when science as practiced today pushes researchers toward projects that will be completed in a matter of months or, at most, a handful of years. As a breach of that rule, conducting a decades-long whole-ecosystem study pales beside what Harmon is attempting.

These logs, although hardly the largest in the old growth, will not finish rotting for two centuries or more—meaning that it will be up to the contemporaries of Harmon's great-great-great-great-grandchildren to complete the analysis he has begun. (If Thoreau had begun such a study in, say, 1850, it would still be a half century away from being completed!)

But although the full story of the fallen log may not reveal itself for centuries, Harmon says his detailed work, and research by the Andrews scientists that preceded his project, has already established that logs are one of the most overlooked of the critical components of the living ecosystem.

Harmon, in fact, goes further. He handed me a chunk of rotting wood the size of a hamburger bun. Although it seemed solid, a closer look revealed that the inner bark was long gone. And much of the sapwood, in fact all of the cellulose, had vanished. The heartwood remained intact, but the sapwood was only a porous mesh: lignin, Harmon said, that was yet to be broken down. But, in time, it, too, would vanish, feeding a long chain of thriving life in the rotting log.

"Somebody made a mistake," he said, with a quick grin. "The tree that's green and standing up is the one they should have called 'dead.' The tree down on the ground is the one that's *really* alive."

Harmon was not being completely frivolous. He can prove that a fallen tree in an advanced stage of rot can hold far more mass of living tissue than a live and standing and apparently thriving one.

Trees die in different ways. That is, they die not only from various causes, but in different fashions. One may be partly consumed over a matter of minutes in a fire. Another might be instantly felled by a down-drafting wind. Another may die slowly from, say, a root-infecting pathogen, and its dead hulk may remain standing for decades. (In fact, tree physiologists have found that often trees die from multiple effects: the "cause" may appear to have been rot, but the rot may have begun with decay fungi ferried into the heart of the tree by invading bark beetles, which themselves entered a wound in a tree caused by a scratching black bear.)

The way in which it perishes has much to do with the role a dead tree will serve in its period of afterlife in the woods. A tree that dies on its feet—that is, remains standing as a "snag"—will contribute its own manner of diversity to the structure of the woods. As it dries and rots, it becomes home for a host of insects that bore into it. Those insects become a source of food for birds, like woodpeckers, that work the tunnels in the cellulose for larvae. The holes the woodpeckers bore might become nesting holes for other bird species, such as nuthatches, or even bats. In fact, a tree needn't die entirely, or immediately, to become what some ecologists have called vertical deadwood habitat. Tops of trees can die as a result of, say, a lightning strike and remain in place even while the rest of the tree continues to grow.

A falling tree triggers a wholly different chain of events in the ecosystem. Especially when a huge, old growth giant crashes to earth, the structural changes in the forest are obvious, and immediate. Where there was deep shade, there now is a gap, and a flood of light reaches the forest floor, triggering the growth of shrubs and young trees. (The gap is likely larger than the path of the fallen tree itself, for it will likely have brought other, smaller trees down with it.) That, in turn, maintains a high level of diversity in the old-growth forest: young trees, and middle-aged trees, grow in gaps among the ancient giants. The height and structure of the canopy thus varies tremendously. The old-growth forest becomes a patchwork of shadow and light. And the woods can provide habitat for a broader range of species.

Other sorts of variations tend to increase diversity. A tree that breaks off near the surface of the ground leaves its roots and stump in place to rot slowly and merge with the soil. A tree that is uprooted exposes bare mineral soil and, perhaps most important, helps form a hummocky topography on the forest floor. Such a sheer variety of landforms, and thus minihabitats, on the forest floor lead to a greater diversity of life-forms.

The old view of fallen logs, says Harmon, is that "the organisms that inhabit them are troublemakers. Logs were seen as harbors for all kinds of bad fungi that rot trees and all kinds of nasty insects that hurt trees."

But his studies, melded with research by the earlier Andrews studies, have begun to show that even the process of rotting itself provides far greater value to the woods than anyone previously imagined. Where only about 5 percent of a live tree might be living cells, he says, as much as 20 percent of the weight of a rotting log can be composed of living tissue. The decomposers that invade a fallen tree seek to break down and consume its carbon, an increasingly difficult task as the decomposers move from recently living tissue to the lignin-impregnated cellulose in the wood, and particularly the decay-resistant heartwood at the core of the tree.

The log I stood on at Reference Stand Number Two had fallen over in the winter. By the following spring, Harmon says, the inner bark will have begun to ferment, producing just a whiff of alcohol that escapes through the bark and excites and attracts whole platoons of ambrosia beetles. Within days, the beetles will have bored their way under the bark, pushing out bits of inner bark and wood as they go. The beetles are not looking for food from the tree itself, but rather for recently dead wood, of just the proper humidity, in which to build a vast network of tiny tunnels. As they arrive, they ferry into the log bits of specialized fungi that cling to a special structure near their heads. The fungi will move from beetle to wood and will begin to decompose and consume the wood, and flourish. The ambrosia beetles, like farmers, will feed on the fungi. But conditions must be perfect. Too little moisture in the log and their fungal crop will fail. Too much moisture and the fungi will clog the very galleries that beetles have dug. If the first disaster strikes, the beetles' larvae will starve. If the second, they will smother in their own proliferating food.

In those same early days of decay, a female bark beetle might discover a crevice in the bark. As she begins digging her own gallery, she will release a signal, a pheromone, that will attract a host of her kind, who will gnaw their way through the thick, tough bark. No longer suppressed by the defenses of the living tree, they will move like a sort of arthropod corps of engineers, making great networks of tunnels. These will allow water and other invading organisms to find their way into the tree, and deeper into the cellulose, over succeeding months, years, decades, even centuries. Indeed, the bark beetles function not just as bulldozers of these roadways, but, like the ambrosia beetles, as taxis for other invaders. The blue-stain fungi that Harmon had shown me in his Douglas fir cookie, for example, had probably found its way into the log on the body of a beetle.

The bark beetles tend not to travel deep. They penetrate just past the nutrient-rich inner bark, boring in order to chew out a gallery in which to lay fresh eggs. Weeks later, when larvae emerge from the eggs, they will bore along the inner bark even farther, eating this carbohydrate-rich food as they go, progressively enlarging a network of food galleries and tunnels. They will enrich this developing new structure in the ecosystem with uneaten shreds of tree from their borings, and pack their galleries with their own nutrient-rich feces.

The succulent inner bark is easiest to digest. And it is gone in a matter of months. A new wave of invaders arrives. The ponderous borer will lay its eggs in cracks and crevices in the decaying wood.

Eventually, its larvae will chew, slowly (yes, ponderously), deep into the very heartwood in a life span that can last seven years, over which time they can grow to two and a half inches in length. Carpenter ants will invade, building more galleries for their eggs.

By its second year of decay, the log is already riddled with thousands upon thousands of holes. Although it is still solid and intact, tunnels snake through its outermost surface, and it is home to hordes of organisms—and not only the wood borers. With the wood-boring adults have entered even other animals, hosts of microscopic mites and nematodes—the roundworms. Predators and parasites follow the first wave of insects, too. Minuscule wasps, for instance, can locate the presence of a wood borer's larva through the bark. (Researchers were puzzled for years about just how the wasp could find a tiny larva covered by tree bark. Odor, they found, was not a factor: wasps still found larva wrapped in a thin and airtight shield. Scientists finally discovered that the bark is just a tiny fraction of a degree warmer over the larva, a whisper of heat that the tiny wasps can detect with their antennae. Once it locates a victim, a wasp will drill its ovipositor through the bark, sting the larva, and deposit its own egg adjacent to it. Soon hatched, the new wasp larva finds itself with a ready supply of food.)

Along with the steady invasion of insects and other invertebrates comes an array of fungi that specialize in breaking down wood, riding the bodies of the tiny animals. The first round of fungi seek out easily available nutrients. But by perhaps the third or fourth year, a new group will enter. (By now, the log is so riddled with holes and cracks that spores of fungi riding on the wind, or in the rains, can simply settle on the log and find their way inside.) These powerful wood rotters begin the difficult task of breaking down some of the toughest lignin-impregnated wood, and set the stage for another sort of animal invader, a veritable ecosystem on six legs.

At perhaps ten years, when the wood is suitably moist, armies of Pacific dampwood termites will enter. Their guts happen to provide a perfect chamber for anaerobic protozoa—organisms that thrive in the absence of oxygen. The termites devour tough cellulose and shuttle it to the protozoa in their guts, which ferment it to provide their own nutrition. The termites, in turn, absorb through their hindgut one of the by-products of fermentation, acetic acid. The insects actually oxidize the acid for the benefit of its energy, in the same way that trees, and humans, metabolize sugar and other carbohydrates for energy. Meanwhile, the termites' guts are filled with nitrogen-fixing bacteria, which process nitrogen from the air and provide it first to the termites and, as they die, to the otherwise nitrogen-poor ecosystem in the log.

Henry Cowles, that dune-wandering first American ecologist, would have appreciated the entire process. Like the successional development of a forest itself, a rotting log moves through successional stages as well. In the tree's death the creeping and crawling invaders, the beetles, the fungi, and other organisms will gradually fill the log with more and more holes, channels, tunnels, more structure: more and more life, a continually changing cast of decomposers that will slowly process and release the log's nutrition back into the forest. In the old tree's life, the bole that would become a fallen log was, by weight and volume, almost entirely made up of dead wood. But paradoxically, when the tree is dead, and the fallen log is thoroughly in decay, it will be positively teeming with life.

After centuries pass, the log may disappear, although it may take as long as two or three or even four centuries of teeming life and steady decay for it to become totally incorporated into the soil.

In retrospect, it is clear that rotting logs are, at the very least, critical parts of the biodiversity—the sheer richness of life—in a natural forest. But even before Mark Harmon began his studies, the Andrews scientists had concluded that logs were critical to the forest in even more direct ways—not that the notion occurred to them in the beginning. According to charter members of the Andrews team, virtually all of them ignored the fallen logs that strewed not only the forest uplands and clogged the swift creeks that flow through the site, but literally saw them as a nuisance.

Botanist and Andrews Forest manager Art McKee laughs now at the memory of the researchers attempting to assess how organic matter—the carbon complexes produced by photosynthesis—moved through the system, during baseline studies in the project's early years. During the early stream surveys, he says, "Here were the guys out with little baggies collecting leaves and twigs from the water. Meanwhile, they were cursing and busting their butts climbing over all these logs that they otherwise weren't paying any attention to."

In McKee's view, there was a simple reason why the logs were initially ignored. "Virtually all of us were trained as ecologists in the East, where there isn't any significant old growth left." Everyone, he says, simply assumed they should use collection and surveying techniques they had learned in eastern forest ecosystems, where masses of old logs simply did not exist.

Just how the researchers made a transition from ignoring logs to paying a great deal of attention to them has been lost in the mist of memory. But McKee says he is certain that the notion that the logs might be key components of the living ecosystem evolved more gradually than Bill Denison's epiphany about nitrogen-fixing lichens in the canopy.

"I don't think there ever was a time when we said, 'Oh wow,'" says McKee. "It was more gradual. But by the mid- to late-1970s, a new picture was coming together."

And that picture had everything to do with an important question the Andrews scientists had begun to ask themselves. If the process of new succession in the forest began after a fire, or a blowdown, or logging or other disturbance, just what factors drove that succession? How did a once-thriving old-growth forest restart itself after a catastrophe?

The Andrews team was to discover that clutter, in the form of standing dead trees, logs on the forest floor, and logs in streams, was a key part of the successional engine of the ecosystem itself. The logs, it turned out, play an especially significant role when the forest must rebound from disturbances.

From completely ignoring the logs, the team eventually would turn its opinion around 180 degrees. "We came to believe that logs were the focus of recovery in the forest," says Fred Swanson.

In general, the Andrews team has determined that old-growth forests are resilient in ways that younger forests are not. Says

Andrews scientist David Perry, "Older forests, because of aspects of their structure, are more resistant to a small fire turning into a large fire." (Perry was referring specifically to the Pacific Northwest. Some forests, like the jack-pine forests of the northern Great Lakes region, are evolutionarily primed to burn as they reach their equivalent of old age.) Small trees, with their less dense trunks, thinner bark, and smaller branches, simply ignite with the application of far less heat. In young forests, branches are closer to ground fires, and the branches are interdigitated in a closed canopy, without the gaps created by fallen trees characteristic of old-growth forests. Thus, in young forests the fire tends to move from tree to tree with far more efficiency.

But the scientists found that logs also are part of the fire resistance. Hardly the dry, dead wood foresters once assumed them to be, the Andrews researchers discovered that they were supersponges. Particularly after all the work of beetles and fungi create a maze of crevices and cracks, they can accumulate enormous volumes of water and can retain it even *after* an intense fire at the peak of a summer drought.

Some logs serve as "nurses," particularly western hemlocks, as well as the giant Sitka spruces that tower in the foggy coastal rain forests. Harmon's research has shown that these trees have trouble seeding themselves on the moss-matted forest floor. But the relatively bare top of a fallen log in its early stages of decay, dappled with bits of humus made from the decay of fallen twigs and other small litter from above, provides an ideal bed on which seedlings can begin to grow. As they decay further, these nurse logs, now teeming in death with nutrient fungi, serve essentially as the now-penetrable and wet "soil" on which the spruces and hemlocks will flourish. (Something similar occurs in the northern hardwood forests of the American East. In the extremely scarce parcels of extant old growth in places like upper Michigan's Sylvania wilderness, young eastern hemlocks tend to thrive, even though elsewhere in Michigan and adjoining Wisconsin the species is in sharp decline. One reason: a quick look at the old logs that litter this forest's floor reveals that hundreds, even thousands, of tiny seedlings will grow along the top of such a nurse. Another reason, however, is probably that the deep old growth is a relatively poor habitat for deer, which tend to thrive in younger, more frequently logged forests, and which favor the tasty shoots of young hemlocks.)

The logs serve, too, as refuges for forest animals. Salamanders use severely rotted logs as habitat. Small mammals, ranging from voles to Townsend chipmunks, take shelter in or around them. The researchers found that logs on mountain slopes tend to accumulate rich deposits of soil on their upslope side, not only providing a rich burrowing habitat for a host of insects and small mammals, but helping to control erosion on even a steep forested mountain slope. On their downslope, the overhang they provide means shelter and nesting spots.

And, says Jerry Franklin, it was not only that the downed logs provided habitat for living organisms. "In terms of nutrients, we found out that downed logs are kind of a neat package. Catastrophes might kill large trees but they don't really consume very much of them, so they become logs that are huge packages of nutrients that are slowly made available to the forest."

Just as striking as the discoveries about the value of fallen logs on land, members of the Andrews team who looked at adjacent streams found that, contrary to then-conventional wisdom, coarse woody debris in streams played an invaluable ecological role as well.

Foresters, according to the Andrews's Art McKee, once considered clearing logs out of streams to be just as much a reasonable practice as clearing logs from land. In fact, he says, in addition to allowing loggers to retrieve marketable logs from streams, the U.S. Forest Service often required loggers buying timber from federal lands to *clean up* logjams and jumbles of big wood in streams, even when not marketable, as a condition of the sale. Among other things, he says, the huge old logs became de facto dams, or partial dams, jamming themselves together in streams. They were therefore considered "impediments to fish movement." As odd as it might strike any stream fisherman (or great blue heron) who knows that it is precisely under and around the most seemingly impassable obstructions that fish like to take cover, the practice of cleaning up streams was common on western federal forest lands for decades.

But the Andrews team discovered that logs in the stream were not only *not* a general impediment to the movement of fish, they played far more important ecological roles than anyone previously imagined. In streams, logs and large limbs alter channels, redirect flows, and force the current to slow down and vary in intensity. In a stream where woody debris has been neatly cleared away, banks and bottoms tend to be uniform and well graded. In contrast, the log-choked waters of the Andrews old growth offer far more complex structure, where regions of fast-flowing waters pour into deeper, still pools and long gentle riffles where creatures like the amphibious Olympic salamander can feed without being swept away by current.

Currents rushing into logjams slow dramatically, dropping their sediments to the stream floor, here and there, providing more habitat variety for creatures like bottom-living insect larvae and other aquatic arthropods. Individual huge logs, and especially the tangled, spectacular jams that look as if entire log cabins had shattered across a stream, slow down the currents, dissipating the flowing water's energy, thus reducing its ability to erode stream banks, which could cloud and clog with sediment an otherwise clear-flowing stream. At the same time, the logs and logjams slow down the flow of nutrients that have entered the stream ecosystem from the surrounding forest, so that they can be consumed in these upstream reaches by such organisms as small invertebrates that form the animal base of stream food-webs. Significantly, researchers have found that salmon reproduce and survive far better in reaches of stream that run through log-choked old growth than they do in the more channelized, cleaned-up streams of commercial plantations. Several species of salmon that should be thriving in the oceans adjoining the Pacific Northwest are either endangered or approaching that status. Virtually all depend on the freshwater streams of the region for reproduction.

In terms of beginning to understand just how important logs are in the stream ecosystem, stream ecologist Stan Gregory says, "It just became increasingly clear that the parklike vision of a stream—just like the parklike vision of the forest—was the antithesis of the natural world. The more we looked at the streams, the more we realized that there were attributes of the undisturbed systems that we just weren't seeing in the managed environment. The most absurd extension of our tendency to channelize and clean up the stream are like those concrete channels you see in Los Angeles. The streams in old growth are filled with wood, and they are tremendously diverse and simply very messy, very irregular places that can change very rapidly."

Discoveries about the critical importance of logs in streams, Gregory says, would not even have been possible in streams running through the second growth, and long-managed woodlands of the East. "Those systems have been so profoundly altered for so long that it would have been impossible to recognize the importance of woody debris." Even during early stream surveys in the Andrews, he says, researchers simply failed to recognize how much of a key role wood might be playing. "There was wood everywhere."

In a more recent development, Gregory has affixed time-lapse cameras along the banks of creeks in and outside the Andrews. As a result, he has been able to watch the streams as they breathe—expanding and flooding the adjoining woods in response to snowmelt high in the Cascades or rainfall, then retracting again.

By the 1990s, he and his collaborators were watching frame-by-frame photos showing immense logjams, which can weigh hundreds of tons, actually bobbing up on the flowing current like so many pieces of cork when waters ran high. "We found whole huge logjams that float up in high waters but would drop right back into place afterward with very little indication that they'd moved at all, even when they had floated up a meter or more," he says.

The discovery is now helping engineers design artificial logjams built in some streams to help restore spawning and rearing habitat for salmon. In the past, such structures, no matter how vigorously anchored into the streambed or surrounding terrain, often simply tore loose and fell apart, to the dismay of fisheries managers. No one understood why, until the Andrews study proved that the answer was the logs' own fantastic propensity toward buoyancy. So Gregory and other members of the "stream team" have begun to help engineers design "smarter" artificial structures that move in response to high waters, just as modern skyscrapers move with earth tremors or high winds.

In the end, says Mark Harmon, "the story of logs in this forest has taught us that there's an incredible invisible world. There's all this stuff going on with the log. There's all this invisible stuff going on *inside* the log. And it's all very integrated with the forest and the way it lives."

6

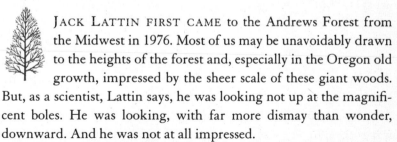JACK LATTIN FIRST CAME to the Andrews Forest from the Midwest in 1976. Most of us may be unavoidably drawn to the heights of the forest and, especially in the Oregon old growth, impressed by the sheer scale of these giant woods. But, as a scientist, Lattin says, he was looking not up at the magnificent boles. He was looking, with far more dismay than wonder, downward. And he was not at all impressed.

Lattin is an entomologist and a taxonomist—a classifier of insects and related arthropods ("jointed feet"), like spiders and mites. When he first came to the Andrews, he said, "I couldn't help comparing it with a [then recent] visit to the University of Michigan biological station [in Michigan's northern lower peninsula]. When you go up there, there's lots of insect activity. You roll over a log and insects come out all over the place. That didn't happen in the Andrews. You rolled over a log and you didn't see very much. I knew there were insects there, but it certainly didn't seem to be a very exciting place—not at first glance anyway.

"Like most first glances," he adds, "it turned out to be wrong."

Lattin tells me this in his lab, the Entomology Laboratory at

Oregon State, which he directs and which is, in a manner of speaking, a taxonomic library of arthropod life. The neatly preserved remains of more than two and half million arthropods are here, most pinned up and neatly labeled in glass trays in hundreds of wide, flat drawers in aluminum cabinets that are stacked to the ceiling in the lab. Lattin, a nationally respected entomologist, is effusively friendly and utterly unassuming. He has bushy, graying eyebrows, swept-back black hair, and a proclivity to white dress shirts, sans tie. He pulls out trays of Hemiptera, the true bugs, to show me the fat-ovoid specimens, with names like *Elasmostethus cruciatus* and *Salignus distinguendus* and *Stictopleurus punctiventris*, and then a sort of gee-whiz tray with an immense, crucified Goliath beetle, a huge Polyphemus moth, a giant walking stick, and an assemblage of similar curiosities of his entomological world.

"Insects," he says, "tend to get overlooked because they're small and—aside from butterflies—people think of them as something to get rid of. But if we took an insect and made it as big as a bird, people would find out they're every bit as nice."

Nice or not, they are the crowning achievement of evolutionary success. If the pessimists are correct—if indeed sentience is a flawed experiment, and we humans are doomed to poison or overpopulate or irradiate ourselves into evolutionary oblivion—it does not mean that the insects will inherit the earth from us. They need not. The war of the numbers was over long ago. The insects already have won.

An intriguing debate among biologists over the last two decades involves the sheer number of species on earth. In his landmark 1979 book *The Sinking Ark*, British ecologist Norman Meyers notes that biologists disagreed about those numbers by a striking margin—the most conservative estimates were that some three million species inhabited the planet. The highest estimates were considered extravagant by many scientists: that the earth was home to some ten million species, although even then Meyers speculated that the "informed guesstimate" of as many as ten million species could indeed be too low. By the 1990s, the estimates had exploded, to the point that *thirty* million species was reasonable.

The expansive new view was based in large part on work by

Smithsonian Institution biologist Terry L. Erwin, who in the late 1970s began fogging a series of tree canopies in tropical rain forests in Central and South America with insecticides. As insects and other arthropods fell to the earth, Erwin simply collected the arthropod fallout and began to sort and count. Based on the sheer numbers of new arthropod species he found, Erwin extrapolated to suggest that there may be at least thirty million species on earth.

Other biologists have insisted that the numbers could not possibly be so high (but a few others have suggested that Erwin could be underestimating). Any debate about the number of species, in fact, is not so much about how many species there are in general, as about how many types of insect exist—or, rather, insects and related arthropods (spiders, ticks, mites). The number of plant species is relatively puny: somewhere under five hundred thousand species, even if one charitably accepts mosses, lichens, liverworts, and fungi as plants. The number of vertebrate animals on earth is strikingly lower, some forty-one thousand.

When a new bird or mammal is discovered, it is major news in biology: like the widely hailed discovery deep in the Amazon rain forest in 1992 by Marco Schwarz, a Swiss biologist, of a heretofore unknown species of monkey, the Maues (pronounced mah-ways) marmoset. By contrast, even though most entomologists agree that far, far too *few* hours and resources are devoted to identifying insect species, some twenty new ones are being discovered and described daily. The Harvard biologist Edward O. Wilson once discovered two hundred new species of ants "within hours" after arriving in a tropical rain forest in Brazil. Among the arthropods identified in the Andrews so far by Lattin and his colleagues, literally thousands are new to science.

Meyers noted that even based on the old estimates of insect numbers, the weight of the one quintillion insects on earth may exceed the combined weight of the human population by a factor of 12. If Erwin's estimates are more accurate, the number, and weight, of insects is roughly four times greater.

Insects and related arthropods have thrived on earth for some three hundred million years. They flourished long before the rise, and then fall, of the dinosaurs. The reasons for their extraordinary evolutionary success appear to be, in part, their small size (some are microscopic), their mobility (as in flight), and, most especially, their short life spans. Mayflies, for instance, are evanescence defined: *Ephemeroptera*. Their adult lives consist of a few hours of flight during which they breed, before spinning, like infinitesimal snowflakes, back to earth. Arthropods have adapted so extraordinarily to so many microniches simply because natural selection among highly mobile and tiny species that live short lives and reproduce so rapidly can occur at what amount to, in evolutionary terms, blazing speeds.

Nevertheless, Jack Lattin's first impressions of the Andrews were hardly out of line with the conventional wisdom. The coniferous forest in general was never considered, for all its majesty, to offer a great deal by way of biological diversity—and *certainly* not compared to the tropics. Ross H. Arnett, Jr., research taxonomist with the Florida State Collection of Arthropods, and Richard L. Jacques, Jr., a professor of biology at Fairleigh Dickinson University, in their 1985 book *Insect Life* offered the conventional view of the conifer woods: "Although its insect fauna is richer than that of the tundra, the diversity of species is not great, because of the limited habitats." The two entomologists suggested that insect life in the conifers was limited to a relative few creatures like the wood-boring beetles and wasp caterpillars.

But as it turns out, it depends where one looks. Lattin and his colleagues were to find that their coniferous forest hides some of its most remarkable biological secrets underfoot. For starters, in the moist soils of the old-growth forests, the Andrews entomologists were to discover an astonishing diversity of arthropods—a largely unexpected diversity that, in fact, now appears to rival that of tropical ecosystems. (Among scientists elsewhere, there were some clues that diversity might be extraordinary in forest soils. Notably, in the mid-1970s British entomologist J. M. Anderson, of Oxford University, pointed out that in some temperate woodland soils, there could be "up to a thousand species of soil animals . . . present in populations exceeding one or two millions per square meter."

In the Andrews, they found more than sheer numbers. They discovered that these tiny life-forms in the soil of the Andrews Forest—and, by extension, probably the temperate forest in general—appear to play one of the most critical roles in the health and survival of the giant forest itself. In time, they even discovered that the insect communities of the forest are so finely tuned to their environment that they serve as a sort of precision barometer: a knowledgeable entomologist might, by simply analyzing the species of tiny organisms in a handful of soil, describe in astonishing detail the ecosystem above. There was never any question that the job of trying to catalog all the insects on just the Andrews site was going to be a truly immense job. Says Lattin, "Studying insects isn't like studying birds and mammals, where the number and kind are finite. You're always aware of numbers when you work with insects. You know you're going to take samples out of the forest where you'll get tens or thousands of individuals. Because they're so small, they work their way into many places." Indeed, many of the soil creatures are exceedingly small—as tiny as one or two one hundredths of an inch long, insects as small or smaller than the period at the end of this sentence.

Lattin and a team of fellow entomologists and students began with traditional collection methods. Many of the methods were as low tech as they could be: the entomologists used simple handheld insect nets (butterfly nets) to sweep the dense vegetation; or emergence traps, essentially cylinders with nylon mesh on top, to capture flying insects just leaving the larval stage and taking wing; or pitfall traps, which are little more than funnels in tin cans set into holes in the ground; or "malaise traps" for daytime insects—these consist merely of baffled sheets of dark netting rigged across a suspected flyway. "Berlese funnels" are vaguely more sophisticated, although a bright ten-year-old could easily rig one up: an aluminum or plastic funnel capped with a lightbulb heat source and locked over a glass jar of alcohol. They work well for insects in soil or leaf litter or crumbled, near-decayed wood; the heat from the bulb dries the detritus, driving the moisture-seeking insects lower, until they fall through a screen into the alcohol. Their most high-tech tool was a small rotating net that continuously swung in circles about six feet off the ground, driven by a one-quarter-horsepower motor.

Trapping soil insects is largely a matter of collecting the soil. The process for separating insects and their cousins from a soil sample is simple. Chitin, the waxy-hard conglomerate that forms an arthropod's shell (which is really an external skeleton, to which all of the internal organs are attached) is soluble in oil. Soil particles are not. So Lattin and his colleagues were able to begin surveying soil samples simply by taking a moistened soil sample, shaking it in a jar with a few drops of cooking oil, and then skimming the arthropods from the surface. Once killed and preserved, the insects were classified, either with the naked eye or a magnifying glass or, often, a microscope—just as Linnaeus did it: genus first, then species.

Lattin and his colleagues had managed to count some thirty-five hundred species of arthropods in the forest soil. Yet he believes that number probably represents somewhat less than half of the estimated species present on the Andrews. According to Lattin, some eight thousand arthropod species probably inhabit the Andrews alone, most of them in the soil.

For a sense of how utterly the insects and other arthropods dominate the diversity of the forest, consider that the Andrews researchers have counted a grand total of 143 mammals, birds, reptiles, and amphibians at the site. Entomologists working on the project have taken nearly two decades to get this far, and yet probably have nearly five thousand more soil arthropods to identify.

To describe and name the yet-unnamed species will take more years, even more decades, says Lattin, in part because of the almost overwhelming numbers of arthropods in even a small sample of soil. The survey so far has shown that the soil under a single square yard of forest can hold as many as two hundred thousand mites from a single suborder of mites, the oribatids, not to mention tens of thousands of other mites, beetles, centipedes, pseudoscorpions, springtails, "microspiders," and other creatures. If Lattin and his colleagues had found nothing else, the discovery of such an extraordinary diversity of arthropod life constituted a major scientific breakthrough—and even a shock. Melody Allen, executive director of the Xerces Society, an invertebrate conservation group, says she found the findings "especially surprising because we think of that kind of diversity as being related to the tropics, not the temperate forests."

Taxonomists specialize. Andy Moldenke specializes in those oribatid mites. Oribatid mites are impossibly tiny—small enough that getting a good look at one can require an electron microscope. I first met Moldenke on a wet Oregon winter day in the cramped virtual walk-in closet that was then his office, across the hall from Lattin's Entomology Laboratory.

On first impression Moldenke comes off as an eccentric scientist right out of central casting. With tousled hair, frowsy mustache, and round, almost childlike eyes, he was, that day, wearing a fabulously rumpled blue corduroy shirt and black canvas sneakers. He is given to creative playfulness: as a way of explicating the astonishing sheer numbers of creatures in the soil, he set out once to calculate the area covered by the average human shoe. Subsequently, for an article in *Wings*, the journal of the Xerces Society, he pointed out that "every time you take a step in a mature Oregon forest, your foot is being supported on the back of 16,000 invertebrates held up by an average total of 120,000 legs. Just think how many creatures it takes to support a single tree." He would, not long after our first meeting, send me a note that he would sign: "Entomologically yours."

"We all have filters when we look at the world," he told me that day. "Most of what I look at when I'm in the forest, most people don't even see. And I can't blame them. I spend most of my time looking at the ground. But seven years ago, when I looked at the ground I was just like everybody else. All I saw was *dirt*."

Science, like most other endeavors, moves in fits and starts. For Moldenke, a critical turning point came in the early 1980s, at a conference about soils at the University of Alberta. It was after the conference, he says, that he began looking at forest soils in a wholly new way. Where soils had once seemed to him to be merely dirt—or at best dirt with some nutrients for trees and other plants in it—he says that the Alberta conference led him on a path toward seeing soil not as an assemblage of inorganic particles, but rather as an amazing substance that was in fact "almost entirely biologic in origin." The soil, in short, was as much a part of the biota of the forest as the birds, the mammals, the shrubs, the trees, or, for that matter, the arthropods.

It was an eclectic conference. "These guys," Moldenke says, "ran the whole gamut—there were soil scientists, ecologists, and taxonomists. The purpose of it was really twofold—to transfer the latest thinking about soils in Europe to Canada and the University of Alberta, and to expose soil managers to biologic thinking, as opposed to simple inert dirt thinking."

The breakthrough in seeing forest soils for what they really are began, he says, in Europe, he thinks because of the greater traditional orientation of European biologists toward natural history—a tradition of focusing on whole organisms and what functions they serve in an ecosystem.

"In Europe most biological scientists are trained as natural historians or grow up in the ambience of natural history. So they're naturally interested in asking, say, 'Well, what does a millipede really do in the forest? Or how many kinds are there? And what do the different kinds do that are different in the forest—and how does the forest depend on the functions of that individual species?' You'd think we'd be asking the same kinds of questions here in North America, but all too often that isn't where our science is focused."

Back in Oregon, Moldenke began looking through a microscope at slides containing minute bits of soil. "I was looking at soil samples from all the way down to the mineral layer and realized that *everything* in the soil was the result of the action of invertebrates—all the shapes and all the chemicals. I began to realize that soil is nothing but the bodies of countless microbes and the bodies and feces of invertebrates. And then I thought, 'If this is true, isn't it important to understand what's really going on with the critters down there?' "

In the years since the Alberta conference, Moldenke has been doing just that—trying to figure out what's really going on in the soil of the Andrews forest and, by extension, the temperate forest in general.

Pick up a handful of topsoil next time you are in a forest. Consider this. At least the way Moldenke has come to see it, the soil is life itself, reconfigured. Each tiny particle is a parcel of something, once living, and now some phase or another of a long, slow process where it is disassembled into the chemical foods that will recycle into more life.

Looked at one way, the soil in your hand is ground up bits of tissue of plants and mosses and fungi, bark and heartwood, leaves and green shoots. Looked at another way, it is, as Moldenke likes to say, "all bug poop." Millions of tiny factories have transmogrified once-living matter into the stuff in your hand. Those factories happen to be the jaws and guts of millions of bugs.

What's going on in the soil of the temperate forest, Moldenke has discovered, is that the moment a bit of leaf or twig, or for that matter the bole of a giant tree, falls to earth, the arthropods become the driving force behind the process of disassembling and recycling it. It is a long, excruciatingly complex process, reminiscent, he says with a chuckle, of a cartoon he once saw of a dung beetle eating the dung of a dung beetle eating the dung of a dung beetle.

In order for, say, the nitrogen bound up in a bit of detritus—a leaf or a twig—to be recycled in an ecosystem, it must be converted from an inorganic form to an organic form.

Free-living bacteria in the soil (or bacteria in the guts of insects and other tiny invertebrates) accomplish this feat. In a key part of the chemical decomposition process, bacteria produce enzymes that allow them to feed on the detritus. But it is only when an animal (an amoeba, a roundworm, or an arthropod) eats the bacteria, and excretes excess organic nitrogen from the meal, that the nitrogen becomes available to plants.

In the lush environment of the tropical rain forest, the process occurs so rapidly that soluble nutrients can be washed away if the forest is clear-cut, or burned. In the more harsh environment of the temperate forest, the climate alternates from wet to dry, from cold to hot. "Life in the temperate forest is tougher for the bacteria," says Moldenke. But, he says, they are still the "engine of the ecosystem." Here, food in general is less abundant, and the "savings account" of organic matter in the soil is more difficult to withdraw. So soil arthropods, the key shredders of organic matter, become the limiting and, he contends, the most crucial factor.

Moldenke came back from Alberta with news of a new technology that allowed him to prove it. He and his students have in recent years begun studying soil and arthropod ecosystems using a technique called thin-section microscopy. The process, originally developed by oil-exploration geologists, essentially allows scientists to remove a core of soil about the size of a soup can from the forest floor, to solidify that soil into a solid substance, and then to slice cross sections of the soil from the sample.

Thin-section microscopy is accomplished by insinuating, in a pressure chamber, a still-liquid epoxy into the carefully removed core of soil. Once the epoxy hardens, the now rocklike soil sample can be sliced with jeweler's saws into exceedingly thin wafers, rock hard, but as thin as this sheet of paper, and then polished smooth for examination under a microscope.

The technique thoroughly preserves the soil *with* its parts in place, from larger bits of partially decayed plant matter to microscopic soil particles. On one such slide, Moldenke showed me the image of what was clearly a needle from a Douglas fir—or so it seemed. On first glance, the needle seemed partly decayed, but still mostly intact. But a closer look showed that it was not intact at all. It was a collection of unconnected fragments, thousands of infinitesimal bits arranged in almost precisely the pattern of the needle. Without moving the needle at all, countless tiny arthropods had swallowed parts of it. In fact, every bit of the needle had been chewed up and swallowed. Bacteria living in the insects' digestive systems had worked furiously for a few hours on the outside of the bit of food, extracting nutrients both for themselves and for the arthropod. And then the remaining cell tissue—constituting most of what had been chewed off in the first place—was repackaged into a tiny pellet of ground-up plant matter, and then excreted almost precisely in place. The needle had been, in short, thoroughly reprocessed through the first stages of decay. And it still looked much like a needle, or at least a needle rendered by a pointalist.

The next stage of decay, lasting only a few minutes, occurs when the surface of the brand-new pellet is colonized by another set of bacteria, exuding different enzymes, freeing more nutrients. "After that the nutrients just sit there, and sit there," says Moldenke, "until some other critter comes along and crunches it up yet again to extract its own nutrient tax and forms another pellet."

Indeed, not long after any bit of vegetation or other detritus falls to the forest floor (or even onto a branch in the canopy) insects descend on it, grinding it up. Remarkably, feeding can occur so fast, and the insect can be so small, that the bits of chewed-up needle are returned virtually in place. An even closer microscopic look at each individual tiny pellet revealed that each is nothing more than plant tissue, chewed, chopped up, and then reassembled. In fact, the largest pellets look for all the world like three-dimensional jigsaw puzzles made out of pieces of plant tissue. "The plant cell is still there," says Moldenke. "It's just the bacteria have gotten out most of the easily available soluble nutrients. Then, as you go down further and look at smaller and smaller pellets, you see pellets where insects have actually crushed the cells to get at more soluble nutrients. At a still later stage, you begin to see that structural components [of the plant tissues] have finally been dissolved."

Nature, Jack Lattin suggests, doesn't develop the kind of astonishing diversity of species his team has found in the Andrews purely by accident. There has to be an evolutionary reason for it, he says. Although no one knows why the organisms of the forest soil are so diverse, Lattin believes it's safe to assume that the diversity itself must be "very, very important in terms of its role in that environment."

Is it possible that without maintaining the diversity of the arthropod community hidden in the soil, we could unwittingly doom the coniferous forest, and with it the towering and long-lived trees? Moldenke shrugs. "I don't know *what* the implications of all that diversity are. Neither does anybody else. And that's the scary part. I guess what concerns us most is that the kinds of aboveground ecosystems that most ecologists have studied in the past are a very small part of what's really out there. Most ecology is based on simple paradigms based on what little *has* been studied. When it comes to forest soil biology, I'm supposed to be some kind of guru—and believe me, I don't know *anything* about it.

"But when you have an awful lot of species, it means almost by definition a great number of processes—thousands of different functions taking place. If we instead continue to manage forests on the basis that the ecosystem is much more simple than it really is, we may be setting ourselves up for a big surprise, and it may not be a nice surprise."

A cherished ecological pursuit is that of finding which factors limit the growth of plant life. For example, in the 1960s, a debate raged about whether controlling phosphorus from detergents in the Lake Erie basin would really slow the explosive growth of algae that was choking the lake, virtually killing the natural aquatic ecosystem through a chain of events that led to nearly total deoxygenation of the waters. Although many scientists believed that stringent phosphorus control was necessary, the detergent industry, which used a great deal of phosphorus to boost the detergent's cleaning power, argued that another plant nutrient—maybe nitrogen, maybe carbon—truly controlled the rate of aquatic plant growth, and that excess phosphorus in the absence of more nitrogen or carbon was harmless. In a simple test that turned out to be one of the classic whole-ecosystem experiments of the century, ecologist David Schindler, working at a small, hourglass-shaped experimental lake in remote northwestern Canada, solved the problem conclusively. He placed a plastic curtain between the halves of the lake and added phosphorus to one half. Algae growth almost immediately exploded in that half: phosphorus had been proved to be the limiting chemical.

Moldenke points out that sulfur is often one of the limiting nutrients in forests. "In studying the nutritional physiology of sowbugs in European forests, they found that one of the species of sowbugs is one of the most important processors of sulfur. The story goes like this: sowbugs eat the litter and the mineral soil. As the food passes through their guts, some is picked up by bacteria in the gut, which change the sulfur in the mineral soil from inorganic to organic. And other critters in soil either chew on feces of sowbugs or feed on bodies or feces of those bugs. So the spider gets its organic sulfur by eating the springtail that eats these feces."

(In the Pacific Northwest, Moldenke adds, sowbugs are the creatures that "transduce the last stage of solid logs into that sort of punky wood that becomes something that isn't really wood and isn't really dirt." The sowbug in Oregon, in other words, is crucial to the process of turning immense and solid logs into components of the soil.) "But if you were just walking through the forest, you'd rightly say, who cares about the sowbug?

"A lot of people have assumed that there's always enough nutrients in the deep layers of the soil to take care of trees forever," says Moldenke. "But now we're coming around to the general conclusion that the vast percentage of nutrients that exist in the forest ecosystem are in fact recycled—that the roots that go down deeply are mainly going down to tap water sources, and, at best, a small amount of some nutrients. But it's the soils at the top, those that are almost entirely biological in origin, that contain most of the nutrients used by the system.

"The real trick in this part of the world is that the landscape is so young, geologically speaking. We've managed the forest so far for only one rotation age—or at most two rotation ages. Maybe there's enough residual nutrition in this nice volcanic soil to get us through one or two rotations. But we might not be able to tell right away if we're really messing something up with intensive management of the forest."

Just how important might the arthropods be to the forest? Moldenke would not think the word *paramount* was an overstatement.

"Everybody's interested these days in this notion of long-term site productivity. I think that once you scrape away the politics and the rhetoric, you're going to get down to the fact that the fauna of the soil is maybe *the* most crucial issue in determining long-term site productivity."

One potential practical benefit of all the diversity of invertebrate life in the forest soil is now on the research horizon: the arthropod communities may be able to serve as an exquisitely tuned gauge of changes in the forest ecosystem.

In 1988, Moldenke began plugging data about the tens of thousands of arthropods collected from dozens of sites into a computer for statistical analysis. He claims he did it because he was worn out on discovery, getting "burned out," as he puts it, on taxonomy, even though he was in a sort of taxonomist's heaven, continually finding hundreds of new and yet unnamed and classified species.

"I wanted to know what it all *meant*. If there's anything useful about the science of ecology, it's to go out and make sense out of the patterns—in this case, to try to make some sense out of all the diversity that's there."

If there was a pattern, he said, that's what would have been meaningful. But was there? Not according to any previous analysis. Part of the problem was that there were so many invertebrates in even a single sample that many species would be represented. If there was a pattern, it was overwhelmed by the sheer numbers.

"When you've identified these hundreds and hundreds of critters in a sample, you're totally swamped by data. It's like trying to look at stars at night. There's so many of them you can't really see a pattern, and when you do see patterns you still can't make much sense out of them."

Moldenke decided to look further anyway, using a multivariate statistical computer program to match the relative abundance of groups in arthropods with various sites—wetter soils, drier stands, young forest, new forest.

"The standard ways ecologists deal with diversity is to turn it into a formula—so I thought, why not tap into that information and try to decode it?" he says.

The results were so surprisingly consistent that he worried that the computer had been misprogrammed. A double check showed that it had not. Previously, in the blizzard of data about hundreds of thousands of little invertebrates, no striking patterns had been apparent. But the computer analysis proved that by statistically analyzing the thousands of tiny arthropods in a tin can full of soil from a site, a researcher could predict with accuracy the condition of the site itself.

"The program ran a correlation coefficient for every single species and every single sample—data from over five hundred soil samples each with hundreds of different kinds of critters and thousands and thousands of individuals. It sort of asked, what aspects of all those samples are similar for species one, species two, and then sort of averaged it up, and then it printed it all up in a nice little diagram. It makes a tree of relatedness. And when I ran it, lo and behold, all of the samples on the same branch of the tree came from the same sort of site—and next-door samples had a lot in common."

All of which might seem like little more than a clever stunt. But in fact the discovery is far more remarkable. For years, botanists have known that they could use plant composition of an ecosystem as a sort of meter of ecological condition and change. Think of it in simple terms: If cactus grows on a site, we all know the site is dry. If, on the contrary, cattails flourish, we know it is wet. In a more detailed way, a botanist can look at, say, two northern hardwood forests, both apparently similar, if one looks only at the trees: basswood, yellow birch, sugar maple, beech. And yet the same trees can grow in what are effectively different ecosystems, within, say, a range of climates or in soils of different acidity, moisture, nitrogen, carbon, and so on. For instance, in a Michigan northern hardwood forest, a botanist might find one beech-maple stand where the ground flora belongs to a community dominated by *Osmorhiza*—sweet cicely, and rife with other species such as downy yellow violet, Canada white violet, blue cohosh, bellwort, wild leeks, and miterwort. On another similar site but with more loamy, fertile soil a botanist would still find blue cohosh, but its plant companions would be dwarf ginseng, wild ginger, and bloodroot. On another slightly drier site, even one that looks the same (at least looking up at the trees), a botanist might find a less lush looking ground flora community dominated by wild lily of the valley—*Maianthemum canadense*—and including the club moss called princess pine (which really does look like a tiny pine tree), Solomon's seal, and a sedge called *Carex deweyana*.

The point of analyzing plant communities has been, in part, to use them to assess ecological changes from site to site—or in the same site over time. In the process, plants—or communities of plants—

become phytometers: virtual barometers that can measure a range of ecosystem conditions. That is important for a host of reasons. For one, ecologists need clear, stable keys to map ecosystems. For another, they need meters to measure changes in ecosystems over time. If, for example, a pollutant such as acid rain or a wholesale ecological change such as global warming is altering an ecosystem steadily, but subtlety, the change may not show up in the condition of trees in the forest until the problem is irreversible.

The importance of Moldenke's discovery is that he can do the same thing with soil samples. But because of the remarkable diversity and sheer numbers of the arthropods in even a small soil sample, his meter—a sort of arthropodometer—may offer a far more precise and detailed view of an ecosystem than even looking at the herb community—a meter capable of even finer resolution than the phytometer.

"When people have looked at it before, they assumed that these tiny critters couldn't possibly define a community. The problem is that there are simply so many of them that if you look for any one critter hard enough you'll find it everywhere. But it turns out that in only some areas is it a dominant part of the community. The pattern isn't obvious. It's only when you sit down and look at relative abundance and at functional groups that you see a pattern."

As a sensitive ecological "probe," the arthropod community could be nothing sort of astonishing. Says Moldenke, "If you went to Andrews Forest and brought me back a handful of dirt, I could tell you what time of year you dug the sample up—all critters have life cycles—the altitude it was taken, the slope face—whether north or south. I could tell you the vegetative cover—whether it was Oregon grape or wood sorrel or something like that. I could tell you the successional stage of the forest—whether it is early, middle, or late—whether it was old growth or not. In some areas, I could tell you what kind of tree was nearby and how far away," he says.

"And it's easy! Anybody could do it with a little bit of training. That's the point—there's just a wealth of information. It's like a kind of cybernetics—there's incredible information encoded in soil arthropod fauna of a community."

As an example, he points to a tiny mite called *Eremaeus*, a common oribatid found worldwide and, he says, that looks somewhat like a red-shelled turtle with a pattern of darker red, puncture-type dots on its shell. The mite is common enough, he says, that in every sample one would likely find at least one. But the mite is found in far greater abundance in old-growth sites that are particularly moist. "In the old growth, we'd find thousands and thousands per square meter. [*Eremaeus* are tiny, only about four hundred to six hundred microns—four tenths to six tenths of a millimeter, about one fiftieth of an inch.] Another oribatid mite, *Eulohmannia*, which looks just like a bright orange-yellow gasoline truck, lives in abundance in the earliest stages of forest succession and in the driest sites.

"Because all of these creatures respond quite rapidly to all these gradients, it gives you enormous analytical power."

And that, he suggests, could have sweeping significance in this human-altered world. Consider, just as one example, the effects of global warming. If indeed the planet is warming in fits and starts, the trees in an ancient forest could well be the last to signal that a change—including a disastrous change—has long been under way. "A tree [in the old-growth forest] doesn't tell you too much about what's happening," says Moldenke. "If you want to monitor change in the environment, the worst thing to look at is an organism that's centuries old. But the arthropod community allows you to look at what's happened over a different time frame—as little as a few years, or even a few months. And you can only do that because you have all that diversity."

By analyzing such characteristics among thousands of arthropod data points, a researcher may be able to monitor changes at a site brought on by not only global warming, but perhaps changes in rain chemistry (as in acid or toxic rain) or herbicide use, with new and remarkable precision. At last report, Moldenke and a small group of entomologists at Oregon State were beginning experiments to use their arthropodometer to try to measure the effects of changes in the forest. What, in other words, could this finely tuned probe in the soil tell the researchers about changes in the soil environment in the months and years after an old-growth tree crashes to the ground?

And how might that compare to the effects of removing the tree? Or what about the effects of burning logged sites or using herbicides?

There are, certainly, more discoveries waiting in the rich, deep soils of the forest, among the remarkable invertebrate fauna that live here. Both Lattin and Moldenke are certain of it, even though they're not sure of all the dimensions.

"We've reached the point where we know just a little bit more about the fauna of the forest soil at the end of the twentieth century than was known at the beginning of the nineteenth," says Moldenke. "Fewer than half the species in the soil even on our one site have ever even been given a scientific name. No one's even looked at them long enough to write down if they're blue with yellow stripes or white with purple polka dots. And if we've never even described these species, how can we hope to begin figuring out what their role in the ecosystem is? But they do have roles. If you believe in the fact that ecologic niches don't tend to overlap, then there's every reason to believe there are vast numbers of processes going on that we simply have not developed the knowledge or the tools to properly interpret."

The lack of knowledge is no small issue. It is certain, in both Moldenke's and Lattin's views, that what Moldenke characterizes as "keystone bugs" live in the soil. If so, these creatures are so critical to the survival of the forest ecosystem itself that, should they vanish, the forest ecosystem itself might eventually collapse. And, given the great black hole that knowledge of the ecology of forest-soil insects presently amounts to, it could happen without foresters or scientists even remotely understanding *why* it happened.

America is continuing to manipulate, through logging, through planting programs, through forest management in general, not only the trees in its forests, but also, by extension, the soils in the forests and, by further extension, the arthropods in the soils. And the manipulation of the ecosystem in such a vacuum of knowledge is something that Moldenke says should concern all of us.

"Whenever you manage for what you don't know is going on, you're sure to come up a cropper in future," he says. "If one or two or a few dozen of these arthropods are linchpins in the forest ecosystem, maybe we'd better think twice before we hit the forests with insecticides. And maybe we better think twice about the consequences of introducing an exotic predator." (That is, a predatory insect that could wipe out defenseless native keystone arthropods in the soil. That, in fact, became in real risk in 1990 when one Oregon mill attempted to begin importing Siberian logs to the Pacific Northwest—logs that might well have carried with them a host of exotic pests. One need only contemplate the level of damage caused in the eastern United States by pests like the spruce budworm, or the beetles that carry Dutch elm disease, or the exotic fungal blight that has all but wiped out the American chestnut, to imagine the potentially horrific consequences for the Northwest, which has so far remained relatively free of exotic introduced organisms.)

In the end, Moldenke emphasizes, the tiny arthropods should matter to others as much as they do to him, that is, to anyone who cares about the forest, whether as a resource for timber and pulp or a wild refuge. "I think the reason the man on the street should be concerned is that when you have an awful lot of species, it means, almost by definition, that a great number of different functions are taking place. If that's the case, but if we instead manage forests on the basis that everything's much more simple, we may be setting ourselves up for a big surprise. If it's a good surprise, well, I supposed we could just kiss the ground when it comes. But if it's a nasty surprise, it sure would be nice to avoid it if possible."

One windy but sunny day, as summer was waning, Andy Moldenke and I found ourselves sitting in the forest on a mossy hillside. Douglas firs waved in the wind overhead. Moldenke plopped himself into a sprawling sit-down on the mossy ground. His bare hands, quickly caked with duff and brown earth, plowed deep into the decades-old remains of what had once been a log—a western red cedar, he said— that had subsumed over the decades into the soil, and had in fact become soil. Extracting from the soil and then breaking up damp clumps of what were once wood, he found a panoply of the creeping and the crawling creatures: in black and red warning stripes (its defense is a strong cyanide) a millipede that is so important to the breakdown of the forest soil that it is a veritable keystone. Called the Harpaphe, it is a grinder of bits of wood and detritus. Moldenke says it is the "Waring blender" of the soil, doing a preliminary grind-up of most, if not all, of the bits of leaf litter and detritus that will subsequently pass through other arthropods' guts on the way to becoming soil.

"You have literally thousands of kinds of animals here in this soil," he said, "and probably all of them are dependent in one way or another on the Harpaphe."

He found a Promecognathus ("Oh wow, look: a muncher"), a long-jawed ground beetle, the only predator of the Harpaphe. Then more critters, including the white grub of a stag beetle and a tiny, aqua-colored millipede that looks like a moving thread. (The color comes from the food in its gut, says Moldenke, for the creature is translucent. It has no need to expend energy to produce pigment for itself so deep in the soil.)

I got into the act and broke open a damp chunk of wood from the damp hole Moldenke had dug. It was riddled with burrows and holes. Moldenke showed me a tiny tunnel filled with "channel frass," the excreta—nothing more than ground-up wood particles, Moldenke tells me—of burrowing beetles. We found centipedes, springtails, and "oh, a little mite," which quickly vanished into the wood.

"Here's a baby trap-door spider," he says, and then, "Whoops, he's gone.

"Six weeks after a log hits the ground," he tells me, "inside the log you'll find several dozens of species of arthropods, and within six months two to three hundred. During the log's life [meaning its decay] you'd find several thousands of different arthropods."

But he pointed out something else too: an almost invisible thread. Laced throughout the decaying wood, like an immense, strange horizontal spiderweb are millions of these long, white threads, which are called hyphae, infinitesimal threads of fungus. Andy Moldenke had not strayed upon an unusual spot. As he likes to point out, the soil of the old forests is really a matrix like this of old, decayed, and decaying logs and other detritus that have been ground, digested, and redigested many times over by the insects. (When he teaches classes in forest ecology at the university, Moldenke likes to write on the blackboard this acronym: BPGT, meaning "Bug Poop Grows Trees.")

But at least as significant, the living soil of these woods is a matted webwork of fungal strands like these: little threads much like those Fred Swanson had seen on the newly devastated slope of Mount St. Helens. Science has long known that fungi play a mutualistic role in the life of some trees. But when they began their studies, the Andrews scientists had little idea how intimately these near-invisible fungal strands are woven into the life and survival of the entire forest ecosystem. Little did anyone suspect that here, in a living soil, functioned an exquisite symbiosis among living tree, living fungus, dead tree, burrowing mammals, and even the host of then-largely-undiscovered insects of the soil. Certainly no one suspected that rodents only the size of a little finger, and buried fungi they dine upon, could be drivers of the towering life in these soaring woods.

7

A FEW DAYS AFTER MOLDENKE HAD ME DIGGING in the soil, I found myself back a few dozen miles west of Corvallis, trudging in a drizzle through wet sword fern up a steep slope in a forest on the flank of a mountain called Mary's Peak, in Oregon's Coast Range. I had come here with another group of scientists from the Andrews team to take a closer look at those very fungi.

Dan Luoma, a mycologist, or fungus specialist (and no relation to this writer), was the first to score.

"Trufffffflllllllllle!" he sang out from up a slope covered with sword fern. And he found the next one, too: "Truffffffllllle!"

Randy Molina was almost as quick, singing the word *truffle* as Luoma had done, and in an instant, mycologist Jane Smith was sing-shouting it out, too.

The three mycologists were in the process of trying to determine whether the species of the truffles they were finding varied from forest to forest, based on its age—whether, indeed, these undersoil fungi proceed through a process of succession just as forests do. "We're looking at three age classes of Douglas firs," Smith told me as we scrambled farther up the slope. "Young trees, twenty to thirty years old. Trees fifty to sixty years old like these. And old-growth trees more than four hundred years old. We're thinking there might be different stages in fungi as well as in the trees themselves." In other words, the scientists were trying to nail down whether there was such a thing as, say, old-growth truffles.

The forest floor was matted with moss and duff underfoot. Molina found a subtle hole in the duff, as if a child's thumb had been inserted. "A squirrel," he said toeing the hole, and then suddenly was on his knees digging with a sort of long garden tongs that he'd also outfitted me with—called a truffle fork.

"Let's see if he left any behind," said Molina. But nothing was left behind, and Molina gingerly replaced the soil.

"Just because you have trouble finding anything doesn't mean there are fewer truffles," said Luoma. "It might mean that there are more hungry little mammals."

Trouble finding anything more did not last for long. The patter of rain was quiet but steady on the leaves of ferns and wood sorrel all around, and we moved farther up the slope.

Molina: "Truffffllle!"

The bit of fungi Molina next dug out of the soil was only about the size of a lima bean. It was a sort of dirty beige, an ordinary-seeming little lump nested in his hand. Dan Luoma pried out the blade of a Swiss Army knife, took the little beige ball from Molina's palm, and sliced the soft lump in half, sniffed it, grinned, passed it to me. A surprisingly strong odor filled up my nostrils: pungent, almost precisely the odor of a raw sea scallop, and not a terribly fresh one.

In Europe, especially, some species of truffles have long been prized as a flavorful, gourmet delicacy. Most famously, in the forests around Périgord, in France, trained pigs and dogs have long sniffed them out as part of a lively, traditional truffle industry. Near the end of the nineteenth century, the king of Prussia commissioned a scientist, one A. B. Hatch, to see if he might be able to outfox [or out truffle-pig] the French—discover a way to grow gourmet truffles as a crop, rather than a hit-and-miss delight in the woods.

But in the process of merely trying to figure out how to domesticate truffles, Hatch made a far more startling discovery. Like mushrooms, which grow above ground, truffles are the fruiting bodies of fungi. Hatch discovered that if he followed the tiny threadlike tendrils that extended from those lumpy fruiting bodies, they were eventually connected to the fine roots of plants, usually trees. Curious, Hatch set up an experiment, growing pine seedlings without these sorts of fungal connections, and comparing them to seedlings with the fungal connections. He discovered that the young trees with the associated fungi grew faster and more vigorously. Hatch dubbed the structures that he found—where fungi and fine root seemed to meld into one another as if they were one organism—mycorrhizae, literally fungus root.

Jim Trappe, a charter member of the Andrews team, now retired, puts it this way: "In the course of doing that, he pretty much figured out the essentials of mycorrhizal symbiosis."

What Trappe means is this: Hatch surmised that somehow the fungi and the tree were each benefiting the other, probably by sharing nutrients. Not only was his surmise correct (although nearly a century would pass before radioactive tracing techniques allowed researchers to prove that the organisms were feeding certain kinds of compounds to each other); it turned out to be the starting point for a series of discoveries that would lead Trappe and colleague Chris Maser to a remarkable ecological story, about a web of symbiotic consequence in the forest ecosystem.

Understanding the essentials of this symbiosis in the Andrews occurred, like many great discoveries, almost by chance. It began one day in the early 1970s when Jerry Franklin, then the Andrews team leader, invited fungus specialist Trappe and Maser, a mammalogist, to visit a parcel of old growth with him.

Trappe admits that during the early Andrews meetings, "I was a little disgusted. I felt that the ecosystem belowground was getting short shrift." But as a direct outgrowth of his meeting with Maser, that was to change, and dramatically. "The result turned out to be far more far-reaching than Jerry, or Chris, or I could have imagined at the time," Trappe says.

During the drive to the old growth, Trappe began telling Maser about his specialty: those underground fungi called truffles. The word itself derives from a Latin root word, meaning nothing more elegant than lump (and the Latin root word was considered vulgar at that). But the lumps that had long fascinated Trappe varied dramatically, in diameters from a few fractions of an inch to the size of a softball, and in colors from white to yellow to blue and even (in Europe) to black. He gave Maser a basic refresher course, too, on the role these fungi play as half of the mycorrhizal partnership with the roots of trees.

Scientists already knew that these fungi, which live deep in the soil, cannot photosynthesize. Instead they survive by enwrapping and comingling with the very cells of fine roots, sometimes growing between the root's cells, and sometimes even penetrating into the cells. Once in place, they survive by extracting food from these fine rootlets, removing sugars made by the leaves high overhead through photosynthesis, as well as certain vitamins.

But as Trappe explained it, the relationship is mutual. When roots and fungi combine in such a way, they acquire their joint name, mycorrhizae, for they work virtually as one, to support both truffle and tree. Fed by the sugar, the fungi begin expanding their webwork of fine strands, the hyphae, which are one cell wide and thinner than a human hair. The web can extend in great tangled mats to cover hundreds of square feet of soil—a reach far more extensive than the tree's root system itself.

The hyphae sponge water from the soil and share it with the tree. They are also far more efficient than the roots themselves at reaching and extracting such vital nutrients as phosphorus from soils. Much of the phosphorus is bound up in carbon-based molecules, which are not soluble in water and therefore cannot be taken into cells. But the hyphae release the enzymes phosphatase and phytase, which liberate the phosphorus and make it soluble.

The truffles share both the water and the nutrients they extract from the soil with the roots and, hence, with the tree itself. It is as if, says Trappe, the roots are specialized fishhooks that can themselves take up and haul in some water and nutrients. But the truffles' fantastic web of hyphae also act like a huge a net, he says, "tapping every nook and cranny of the soil." Mycologists now believe that mycorrhizal fungi effectively connect trees with as much as one thousand times more soil area than the roots themselves.

Even as they pump water and mineral nutrients to the roots, the fungi form a sort of protective armor against disease bacteria around the roots, and sometimes actually inoculate the soil with antibiotics that kill disease bacteria. Studies have shown that many soils that are commonly used in tree nurseries are rife with tiny organisms of disease—pathogens. The same pathogens were either nonexistent, or far less abundant, in fungi-rich soils in the wild.

The fungi even generate growth hormones that can stimulate the tree's own feeder roots to branch out. And more recent research has shown that, with protection from mycorrhizal fungi, trees can even grow in seemingly impossible places. For instance, fir seedlings can grow even in the acid-bathed slag heaps of mines if their roots are first inoculated with a sort of bodyguard, in the form of an acid-protective mycorrhizal fungus called *Pisolithus arhizus*.

When they reached the parcel of old-growth forest, called Wheeler Creek Natural Area, in Oregon's Wheeler County, Trappe began showing Maser little pits in the ground where voles or squirrels had been digging for truffles.

"Chris said he had seen the pits before. He knew small mammals were digging, but had no idea what for," says Trappe.

Intrigued, Maser, the mammalogist, mentioned that he had already concluded that many of the small mammals he had been studying—voles, squirrels, chipmunks, and others—were eating fungi, and perhaps a great deal of it. Indeed, just the previous week he had captured an endangered red-backed vole with a chunk of fungi in its mouth. Now he wondered if the fungi they were eating could be these subterranean mycorrhizal truffles.

A few days later, back at his laboratory at Oregon State, Trappe received a package from Maser. When he unwrapped the package, he found test tubes bearing the pickled stomach contents of a small rodent, the mouselike but tailless California red-backed vole. With a pair of tweezers, Trappe extracted a small sample of the contents, examined it under a microscope, and instantly recognized that the stomach was loaded with tiny serrated and ridged fungal spores. After examination of several stomachs, it became increasingly clear that the voles' diets, in fact, consisted largely of truffles, and that in fact they digested the truffles with such thorough efficiency that most of the waste material left was simply a mass of spores. Virtually every one of the eighty-one California red-backed voles the two eventually studied had stomachs loaded with truffle residues. And, it turned out, there was good ecological reason for that.

Trappe and Maser soon were able to establish that other small rodents depended on the truffles, too, including nocturnal flying squirrels that would glide down from the canopy by night for a fungal meal, as well as various species of chipmunks, squirrels, and mice. Even normally insectivorous shrews carried bits of truffle in their stomachs, as did the occasional pika, and even rabbits.

The two biologists quickly pieced together the story of the newly discovered version of symbiosis in nature. Just as the trees and the mycorrhizal fungi depend on each other for survival, so, too, are the giant tree, the truffles, and the small rodents inextricably linked to the life of the other. And not only that: the tree in death, the rotting log on the forest floor, seemed to be as key a player in this complex symphony as any of the living organisms.

The truffles' fungal mats feed water and minerals to the trees. The trees, meanwhile, feed photosynthate and vitamins to the truffles. As the fungal mats grow and spread, the trees grow and thrive. These thriving giant forest trees thus can provide habitat for some of the rodents: flying squirrels high in the canopy. Meanwhile, when the trees die and crash to earth and rot, they provide vital habitat for tiny mammals like the red-backed voles. The rotting logs, meanwhile, provide nutrients to the spreading hyphae. As they dig in rotten logs and in the soils of the old-growth forest, eating truffles as they go, the voles and other rodents distribute the truffles' reproductive spores via their fecal pellets, either directly adjacent to tree roots as they dig for food, or as rain washes away and spreads the tiny spores from the "latrines" near their nests. In fact, since the log must proceed through its succession of rot for the voles eventually to find suitable soft and punky wood for habitat, even such insects as termites are key players in the symphony as well.

Trappe and Maser were among the first of the Andrews group to begin questioning the wisdom of allowing, even encouraging, timber companies to remove all the logs from the forest floor. "The more we thought about it," Trappe told me, "the more it opened up questions, such as: what happens under intensive forest utilization when we no longer permit this material to decay into the soil?"

For most of us, the only experience we ever have of the roots of trees is when their fierce affinity to water forces them through crevices into our sewer pipes, or maybe when we get a glimpse of their tangle at the end up a blown-down tree.

But about half of any living tree is unseen, underground, a growing and branching and reaching system just like the branch network of the canopy above. The roots, in fact, grow in close proportion to the growth of branches and stem and, in particular, leaves, above ground. The relationship is almost immediate in young trees—strip the leaves from young seedlings, and roots stop growing. It is not as immediately apparent with larger trees. In the spring, roots on some large trees can begin growing before the leaves appear. The response seems to be a reaction to the heat of the soil: studies in a frigid New England winter have shown that roots will grow year-round if the roots are kept heated above about 38°F. The reason for the disparity between young and old trees is that the balance between photosynthesizing leaves and roots on the frail and young tree is more immediate—the two are close together, and the roots use quickly what photosynthate they can extract from the leaves. But on a larger, more mature tree, leaves are farther away, the relationship less instantaneous. Simply for purposes of buffering the food supply for roots, there must be storage areas in the big limbs and the trunk. The roots of the large tree can grow from the roots', and the tree's, substantial store of reserved sugars.

The root tip's affinity for water is preternatural. Fine hairs extending from the root appear almost spontaneously at the slightest whiff of moisture; if the soil becomes completely dry, they virtually melt away. The root, meanwhile, extends itself onward through the soil in search of more moisture. The cap at the tip of the root is scaly and hard, like a helmet—or a thimble. It is the most industrious piece of the tree, probing nooks and nicks and crannies, worming between tightly packed stones (and even the small cracks in sidewalks and driveways), pursuing whispers of moisture all the while, snaking around obstructions, and even, when necessary, lifting great weights. The growing, expanding root of a curbside tree can hoist a slab of concrete or split and lift up more than a ton of solid rock.

The root takes in water through simple osmosis. Consider a familiar biochemistry experiment. Fill a small sack of cellophane (the real thing, not plastic wrap), a product of plant cells, with sugar water. Tie the cellophane around the base of a glass tube, and insert the cellophane water balloon into a beaker of distilled water. Because water molecules are always in motion, and because nature hates a tilt in chemical equilibrium, the water molecules from the beaker will inevitably work their way through the microscopic spaces between the cells in the cellophane. The water molecules will move into the cellophane sack, compelled to try to dilute the strong sugar solution to balance it with the water outside the sack. The attempt is in vain, since the sugar molecules are too large to diffuse out of the sack, but the increased volume of water in the sack will create pressure, and, as if to prove that more water has arrived, the sugar solution will move up the tube.

The cellophane acts just like a living cell wall. And it is exactly by that kind of osmosis through the cell walls of root hairs that the tree obtains its water. The roots already have captured and are rich in dissolved nutrients, achieved not by the movement of water but through a process called active transport, which can select necessary elements—for instance, the nutrient phosphorus—and shuttle them into the root, essentially passing them from molecule to molecule, a sort of biochemical bucket brigade. Nutrient ions thus can travel through the otherwise impermeable cell wall and into the protoplasm.

The fine root, then, is much like a sort of cellophane bag. Like the sugar in solution in the beaker, the solution of water and dissolved minerals in the root tip begs to be diluted by pure water from outside the root. Through osmosis water flows, molecule by molecule, into the tip. Through active transport nutrients enter the roots as well, to be dissolved in the water and drawn, as a nutrient-rich sap, through the length of the finer roots (the same osmotic pressure that forced the sugar water up the tube), up the plumbing of the larger woodier roots and farther up the plumbing of the tree itself. Along the way, the growing tissues of roots, then trunk and bole and bough, use up some of the nutrient-rich sap. But, at last, the remaining sap is pushed (and mostly) pulled all the way to the leaves.

Studying the behavior of tree roots has never been an easy proposition. It was a conundrum at least as difficult as Bill Denison's early attempts to explore the canopy. As Andrews botanist Art McKee explains it, researchers were frustrated for years by the problem.

The problem is twofold: it *is* possible to get a look at the overall structure of a root system by blasting away soil with water under pressure. But by the time the mud is cleared away, most or all of the finest root hairs are destroyed. Conversely, a scientist can carefully, like an archaeologist unearthing a piece of fragile pottery, pick and brush soil away from a smaller section of roots to study, in close detail, just that bit of the system, including the finest rootlets. In neither case is a scientist able to study an entire system. Even when a tree has toppled, apparently unearthing its massive root system with it, much of the spectacular system has been severed and remains underground. And studying a dead root system is hardly the way to analyze how the roots actually grow or otherwise function in the soil itself.

But answers are beginning to emerge. At the University of Michigan Biological Station, in the northern reaches of that state, scientists Kurt Pregitzer, Ronald Hendrick, and Robert Fogel finally managed to conduct studies on roots in the early 1990s in a new sort of laboratory. Not far from the station's cluster of aboveground labs and dormitories, in a mixed hardwood forest, down a set of concrete stairs, past a room full of glowing gauges, is buried the station's "soil biotron," which is really not much more than a capacious concrete tunnel. About one hundred yards long, the tunnel, brightly lit, is lined with huge, shuttered picture windows. The shutters usually remain closed so that light does not affect the undersoil world beyond. But once opened, the windows look out on the soil profile below the surface of the mixed hardwood forest overhead. The windows are plastered with masses of fine tree roots. On a routine basis, technicians position close-focusing video cameras on sectors of the viewing windows to record changes. Thanks to the biotron, Pregitzer and his colleagues have been able to experimentally manipulate the underground world, dosing portions of soil with extra water, or water plus nitrogen. Among other things, they have found that a mere twenty-day dose of extra water and nitrogen caused roots to grow to nearly two and a half times their previous length. In another twenty days, the roots had grown more than 800 percent!

At a U.S. Forest Service laboratory in Rhinelander, Wisconsin, tree physiologist Mark Coleman has used "minirhizotrons," steel and Plexiglas tubes outfitted with magnifying video cameras, that he can lower directly into a small hole in the forest soil. Speeding through videotapes he had made from the rhizotrons, Coleman has been able to observe an extraordinary dance of fine roots that almost instantly appear and melt away rapidly in the soil.

"I thought I'd be able to watch a root extend itself through the soil, slowly, over time," he says. "But it's more like this: one day, there's nothing there. The next day, there's a root all the way across the screen. I see them for a day or two, and then it's like they've done their thing, and they're suddenly gone." (The roots do not, of course, simply vanish. They are eaten by microbes and small arthropods, and thus cycle their rich nutrients back into the hidden underground ecosystem.) At his Wisconsin laboratory Coleman ran a videotape for me. It was the image from a high-resolution magnifying camera running down the side of a rhizotron tube a fraction of an inch at a time. Roots, strands of fungus, grains of sandy soil, even worm tunnels in cross section appeared.

"We've learned a few things," he says. "But what we still don't really understand is how a root is born, how it grows, and how it senesces [dies]. It makes you wonder if the Heisenberg uncertainty principle shouldn't have been applied to our work, rather than physics."

Yet other new studies continue to yield new knowledge. Scientists injecting radioisotopes into small trees have found what appear to be direct links between individual roots and individual leaves on some trees, with a root apparently responsible for procuring water and nutrients for a specific leaf, and the leaf, vice versa, responsible for feeding packaged energy in the form of sugar to a specific root.

And it *is* known that the root systems of trees vary: just as the branching systems aboveground are characteristic of a tree (a vaselike elm, after all, takes on a shape strikingly different from the pyramid, Christmas-tree shape of a fir), so too are its belowground branching systems.

As a consequence of the difficulty of studying tree roots, no one has ever succeeded even in counting the number of separate branching stems or, most certainly, the tiny root hairs that grow just below the root tip and do most of the work of water and nutrient uptake. However, the calculating feat once was performed for one kind of root-rich grass—a variety of rye. On a single grass plant grown under laboratory conditions, researchers painstakingly removed all the soil, counted some six thousand *miles* of root hairs, and estimated fourteen *billion* individual root hairs.

Following in Trappe's and Maser's footsteps, ecologist David Perry came to believe that mycorrhizal fungi are keys to how a damaged belowground ecosystem in the forest—and then the forest itself—heals. Perry sees it as a matter of "bootstrapping." That is, he suggests that the fungi help a devastated forest pull itself up by its own biological bootstraps. As he and Andrews scientist Mike Amaranthus have written, "The picture now emerging is that, at least in some cases, the coupling of plants and soil is both intimate and vulnerable."

And it is intimate and vulnerable even beyond the relationships between trees and mycorrhizal fungi, for it turns out that just as the photosynthesizing trees and other green terrestrial plants feed sugars and starches to their fungal partners, they also positively flood the soil around their roots with this rich, energy-packed food, feeding an astonishing network of other organisms.

"From aboveground," says Perry, "grasses, shrubs, trees are interacting with the soils, feeding organisms that live in the soil. Meanwhile, all those organisms in the soil are doing work, pumping nutrients back into the ecosystem."

Scientists now believe that somewhere around 40 percent of the total photosynthate made by the leaves of trees actually does not feed the plant at all but rather seeps out of the roots to feed mycorrhizal fungi and the rest of the ecosystem that thrives in a special region just a fraction of an inch beyond the roots, a biologically teeming zone called the rhizosphere. Here, according to Amaranthus, "a single gram of forest soil ... about a thimbleful, may contain a hundred million bacteria, and several miles of fungal hyphae."

Perry and Amaranthus believe that this hidden world of microscopic organisms is part of a feedback system that helps the forest to regenerate itself when a fire or other disturbance damages or kills the trees above. In essence, they say, the system appears to have evolved in the presence of fires that leave some trees alive and photosynthesizing. That, and the quick growth of pioneering early successional plants that colonize a site after a fire, continues to provide food and support for the underground matrix of life.

But what happens when a disturbance is so devastating that this hidden, undersoil ecosystem's energy supply is wholly removed, and for a long time? According to Perry, in at least some kinds of forests, the rich fungal and bacterial ecosystem declines. In one experiment in a patch of Oregon forest called Cedar Camp, Perry and Amaranthus provided compelling evidence that this notion is correct. A once-thriving site of about forty acres was completely leveled by loggers in 1968. It has been replanted with new seedlings four times. But most new seedlings died each time they were planted, and the few that survived hung to life only perilously.

From a nearby patch of forest dominated by white fir, the scientists removed from the root zones of healthy trees a small amount of soil, placing a mere 150 milliliters (about the amount of a small glass of breakfast juice) in each planting hole for the fifth round of seedlings planted in the unfortunate site. In just the first year, seedling growth and survival increased by half. Three years later, *only* seedlings that had been inoculated with soil from the healthy, still-intact site survived.

Perry suggests that the problem can be particularly acute in drier areas. The fungi and bacteria appear in the root zone of the soil, exuding de facto glues (in the form of organic polysaccharides, themselves the offspring of the photosynthate made by the trees above) that bind tiny grains of soil into larger, more porous aggregated clumps. This in turn builds a looser soil, less like dense clay and more like the soil one might find in a prime farm field, with far better retention and movement of air and, especially, water through the soil.

With an electron microscope, the researchers were able to examine particles of soil from both the devastated clear-cut site and the adjacent site of thriving white fir. From the healthy stand, the soil particles were riddled with pores of various sizes, meaning that far more surface area would be exposed to water and air. From the damaged site, particles were much like beach sand, virtually devoid of pores.

Studies from Africa also suggest that this same problem—change in the structure of soils—appears to be a key to desertification on once-thriving lands. Perry and Amaranthus suggest that this problem also may explain why so much of the acreage of the once-grassy and -lush rangeland of the American West has become functional desert. (There, tens of millions of acres have been decimated since the introduction of trampling, plant tearing, cattle, and sheep.) The nub of the problem is that these extracellular polysaccharides themselves become food for microbes. Obviously, once consumed they cannot do their job. Thus the soil structure depends on a steady seeping of these substances from the roots into the rhizosphere: without the photosynthate from the plants above, the thriving microecosystem in the rhizosphere cannot survive. And, in a sort of cascading disaster, soil structure itself eventually begins to collapse.

The fungi play other roles as well. Saprotrophs—fungi or bacteria that live on and help decay dead organic matter—have long been well known to science as keystone organisms that, in the process of providing for their own survival needs, break down and recycle nutrients into the ecosystem. Mycologists have recently found that the mycorrhizal fungi also produce enzymes that, for instance, break down organic nitrogen, which cannot be reused directly by plants, into forms that can be used. The fungi also appear to be able to mine molecules of critical, but otherwise immobile, metals that trees need in tiny amounts in order to thrive. For instance, some mycorrhizal fungi exude chemicals called siderophores that serve as chelating agents, releasing iron, a critical nutrient that trees need, in tiny amounts, to survive, just as human do. Beyond that, the extensive mats of fungi that extend through the forest soil seem to serve as storage vessels for some nutrients that otherwise would be likely to simply dissolve in water and leach away. As Molina and Amaranthus once explained it, mycorrhizae and other organisms in the rhizosphere "form a web to capture and assimilate nitrogen and other nutrients into complex organic compounds and then slowly release them into the forest ecosystem."

Beyond giving biological aid to the individual tree, mycorrhizal fungi now also appear to play a key role in the ecological succession of the entire forest over time. Laboratory studies in the 1980s began to show that some species of trees will share and exchange nutrients, using strands of interconnected fungi as a "hyphal bridge." Molina says this suggests that as a forest develops through its inevitable successional stages, the shade-tolerant species—such trees as western hemlock in the old growth of Oregon, or sugar maple in the forests of northern New England—can thrive in deep shade in part because they can borrow photosynthate across this hyphal bridge, from the very overstory trees that are making the shade (and doing most of the active photosynthesis in the system).

Studies of a type of birch in the United Kingdom showed that some kinds of mycorrhizal fungi interact with the roots of trees in young stands, and other kinds interact with those in old stands. The group of fungi associated with older trees seemed better adapted to soils rich in the kind of abundant decaying organic matter mature forests would produce.

In fact, it was to find out more about this phenomenon, which also appears to hold true in the Northwest, that Molina and company were in the woods looking for truffles the day I joined them, and singing out at the discovery of each one. They have already found, says Molina, "several species specific to old growth. And we believe that means that some kinds of fungi would be endangered if all the old growth disappeared."

Ultimately, it all appears to mean that a mature forest is a legacy of its predecessors—indeed that old growth thrives especially because early woody plants (starting with shrubs in the opening created by disturbance) helped set up a healthy rhizosphere community in the soil for what would come decades, or centuries, later. If so, the converse is also true: that failing to account for the need of the site to maintain or reinvigorate its living soil could be, over the long term, a prescription for the kind of poor regeneration that Perry and Amaranthus found at Cedar Camp.

At the same time, knowing more about which types of fungi needed to be where in the woods at which point in forest succession could mean dramatic improvements in scientists' and foresters' abilities to help devastated forest ecosystems begin to recover. It could mean something as simple as inoculating the soil with the proper fungi. Indeed, largely as a result of the discoveries by the Andrews researchers, nurseries in the Northwest now routinely inoculate fungi into the soils of seedlings to help them grow and survive.

The fungi—the truffles and their cousins the mushrooms (many of which are also mycorrhizal)—appear to hold another kind of promise. Montana mycologist Larry Evans is among those who have suggested forcefully that in the fungal diversity of America's forest soil lie great economic riches in the forms of living, and constantly renewable, resources that just happen to be delectable. Italy's white truffles, he recently wrote, sell from up to one thousand dollars per pound. Leading epicures, like the late James Beard, have suggested that Oregon's white truffles were "at least as good" as Italy's renowned varieties. According to a calculation by the North American Truffling Society, a *single acre* of some conifer forests could yield about six hundred pounds of truffles each year—or about a quarter million dollars' worth. Mushroom hunters, in fact, are already increasingly active in the woods. In Oregon and Washington, the 1993 commercial fungus harvest was worth an estimated forty million dollars.

At the end of one of my visits to the Andrews, a group of river researchers made my day by presenting me with a fork and plate of chanterelles, sautéed exquisitely. They had been picked from a place near Andrews headquarters (or so they said—no one was about to tell me exactly where that spot was). The Northwest forests are rich in other prized species, like the delectable morel and the pine-mushroom, and even the matsutake, tinged with the taste of cinnamon, and prized almost beyond reason in Japan, where particularly large, one-pound matsutake could be worth six hundred dollars on the open market.

Indeed, Molina believes that in some forests, the market value of fungi alone that could be harvested year in and year out could be greater than the lumber cleared from a forest every several decades (although he says that would probably only apply to areas where trees don't grow as vigorously as in prime timber areas).

Reason suggests that a modest mushroom or truffle harvest should have little effect on trees or the ecosystem, since truffles and mushrooms are only the fruits of an organism that produces far more fruit than it needs to reproduce, as well as millions of excess spores. Most of the living mass of any of these individual fungi is the interconnected mat of hyphae, collectively called the fungal mycelium. Indeed, preliminary studies suggest little observable harm to forests from picking truffles and mushrooms. In fact, a seven-year-long study, by the Oregon Mycological Association, concluded in 1996 that mushrooms tended to flourish *more* in areas targeted for prime timber harvest than in subprime areas. But Molina says caution is in order, suggesting that science still knows far too little about the long-term effects of fungus picking to be certain that a boom in commercial harvesting will not harm trees, and the ecosystem as a whole.

Meanwhile, Jim Trappe wonders about the global implications of discoveries about the importance of maintaining the fungal health of soils. He points especially to places like Japan and Europe, where some forests are in profound decline—with dramatically reduced rates of growth or outright wholesale death of swaths of forest. Those problems have been blamed on pollution. But Trappe wonders if the problem is less the pollution, and more that "these soils have not had their input of woody debris (and associated development of the right fungal balance). What are the implications of so totally changing the forest soil? The answer is: right now, we just don't know."

(There are, however, some troubling indications that Trappe may be all too much on the mark. Forests in much of central and eastern Europe have been in a state of decline for at least two decades. Some scientists believe that decline can be linked to high levels of acid rain in the region. Dutch scientists have already discovered that the only discernible difference in troubled regions between trees such as beech and spruce that are thriving and those that are dying is the presence of mycorrhizal fungi on the roots. Like the spruce thriving on the acidic slag heaps of mines, the trees with roots packed with mycorrhizae may have fungi-mediated defenses against the acids flowing from the rain into the soil, or at least better resistance to toxic minerals such as aluminum that are mobilized by the acids.)

With the discoveries of both the fantastic biodiversity of insects in the soil and the fantastic role of the organisms in the root zone, the subterranean forest is no longer getting the short shrift that Jim Trappe once was so annoyed about. These days, Jerry Franklin says that, aside from the mysteries of the canopy, all the "big black holes" in terms of understanding the forest all lie in the soil.

As a scientist, Molina meanwhile speaks of "learning to live with uncertainty—learning to deal with the fact that there's so much we don't know, that we aren't able to predict. We need to let people know that sometimes a best guess is the best we can do. The important thing is to guess the answer to this question as accurately as possible: how many relationships can you break until things begin to unravel?"

8

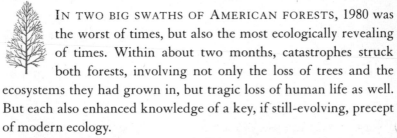

IN TWO BIG SWATHS OF AMERICAN FORESTS, 1980 was the worst of times, but also the most ecologically revealing of times. Within about two months, catastrophes struck both forests, involving not only the loss of trees and the ecosystems they had grown in, but tragic loss of human life as well. But each also enhanced knowledge of a key, if still-evolving, precept of modern ecology.

That precept: ecosystems, and the species within them, are not only resilient in the face of profound damage, but have often evolved to take advantage of it.

Henry Cowles, that first American ecologist, and those like Frederick Clements, who directly followed him, were wrong about one key ecological truth. They were not wrong that ecosystems constitute a web of interdependent life. Their intellectual descendants, such as Barry Commoner, were not wrong to point out that "everything is connected." Especially, they were not wrong that an ecosystem developing on new ground proceeds through an orderly, and generally predictable, pattern of succession—whether it be a newly made and barren dune, or a scoured slab of bedrock exposed as a glacier retreats, or a forest burned or chainsawed into apparent nothingness.

But continuing scientific investigation has proven them at least largely wrong to imagine that succession would lead ultimately to a permanent, self-maintaining, and stable climax community—to the Platonic ideal of a flawless and ultimately fixed order in nature.

Ecologists have come to believe that forest ecosystems, in fact, are as much about disturbance as they are about stability. And although it might seem paradoxical, the rich, even explosive diversity of life in a forest landscape often even *relies* on what can seem like catastrophe to the human eye, or the human spirit, simply because the weaving and wedging of evolution have altered species' very behavior to respond to their environment. If the natural patterns of change in a species's environment is a catastrophe (or a catastrophe as we see it), then a species might even evolve in such a way that it cannot survive *without* such a catastrophe. At the very least, evidence now abounds that many kinds of ecosystems can only be reinvigorated and renewed when visited by a major disturbance. The concept applies, for instance, to the galaxy of tiny "pothole" wetlands that dot the prairie of north-central United States and central Canada. These wetlands, which are crucial for the reproduction of dozens of species of ducks, as well as such waterbirds as white pelicans, must periodically experience drought so profound that they go dry, or nearly so. Without drought, some of their aquatic plant species cannot germinate and these wetlands cannot thrive ecologically over the long term. Paradoxically, the drying of the wetlands may kill vast numbers of the birds that depend on them for survival.

The same rule applies to many types of forest, which must, like the phoenix, be devoured by flames (or some analogue of flames) in order to be reborn. No forested region in the world demonstrates this concept more directly, simply, and poignantly than a spot half a continent away from the Andrews, in the northern reaches of Michigan's Lower Peninsula.

The May 1980 fire in that place was to have been routine: a controlled burn-off of jack-pine cones and needles and bits of scrap and duff that remained on three hundred acres of recently clear-cut forest. Such controlled burns are common—and in fact virtually mandated by both common sense and the particulars of federal endangered species law—here on a sandy glacial outwash plain in the Huron National Forest. Still, on that day, foresters would have preferred to wait for wet weather and low winds to minimize risks and keep the fire as low and cool as possible. But in a spot that is, curiously, just about the coldest (in winter) and yet has some of the hottest recorded summer temperatures on the mitt-shaped Lower Peninsula, it had been an unusually dry, hot spring, making a controlled burn virtually impossible. After waiting day after scorching day, workers from the national forest finally got the go-ahead from their district supervisors on a spring morning that promised to remain calm.

The burn started normally enough, the workers sizzling off dry twigs and needles and ground vegetation with drip torches, which work like weak, miniflamethrowers and allow some control over the pace and direction of the burn. But then the wind kicked up suddenly and blew a spray of sparks over a state highway and into the desiccated litter on the standing, bone-dry jack-pine woods beyond.

The great conflagration that followed would come to be known as the Mack Lake fire. Jack-pine stands are naturally prone to burn vigorously, their dry needles building up for decades beneath them, virtually waiting for a fire to release their nutrients into the soil. In fact, jack pines are so adapted to fire that their waxy cones normally will not open and set seed unless exposed to high heat. Flames had been suppressed for decades in the Mack Lake area. Now these resinous woods exploded, flames roaring through the woods. In its single holocaust afternoon, the Mack Lake fire leveled twenty-four thousand acres. Forest ecologist Dave Cleland would later tell me the fire released the equivalent of "nine Hiroshimas" of energy.

The fire was clearly a horror in human terms: it entrapped, and killed, one Forest Service firefighter. It mowed down an enormous swath of trees and threatened to push its way into nearby villages and towns.

And yet the fire accomplished on a huge scale—and in spades—what it had meant to do on a smaller scale. It actually repaired a large tract of long-troubled ecosystem. In doing so, it helped mightily in the fight to preserve a small gray and white and yellow bird, with a sweet and piercing song, called the Kirtland's warbler. The warbler—a.k.a. the bird of fire—also happens to be one of the most endangered species in the United States.

A decade and a half after the Mack Lake fire, I stood in what had been the heart of the burn with two Forest Service wildlife biologists, Rex Ennis and Phil Huber, and with ecologist Cleland. Our feet amid low blueberry and sweetfern, the fern's odor of spice wafting up, we were surrounded not by the devastation of 1980, but by a lush young forest of jack pines the size of large Christmas trees, some now up to fifteen feet tall. We were listening, in fact, to what would be the only indication that Kirtland's warblers were present. It was past nesting season, and hearing a song, much less finding a bird, was unlikely. But nevertheless we got a single round of a lilting, staccato song, and then silence.

The short version of what has happened here: in the wake of the terrible fire, an expanse of new, young, jack-pine forest has grown up. This, in turn, has provided a vastly improved habitat for the warbler, a creature that evolved here in the presence of some of the most frequent natural fires of any forest ecosystem on the continent (including other jack-pine forests—which are common only in the far northern reaches of the Great Lakes states and parts of Canada).

It was Cleland, who began his career as an industrial chemist, and then became a soil scientist, and then pursued a Ph.D. in ecology, who figured out just why the warbler nests here, or, as he puts it, "why the Dickey bird lives where it lives."

Cleland is lean and mustachioed and, complete with dangling cigarette, could double for Charles Bronson. He is quick with gentle, self-deprecating quips. (High-energy to the extreme, he suggests that he has been presented with promotions and awards because his superiors "keep mistaking metabolism for intelligence.")

On one level, science long has known why the warbler haunts

jack-pine forests. The small bird is a ground-nester. It will only nest under low, ground-hugging—or virtually ground-touching—branches of jack pines (and jack pines only). Furthermore, to find forage for its young, the warbler must have a source of certain kinds of insects that associate with jack pines and, especially, a matrix of sunlit openings where blueberries and other easily picked and nutritious edibles will spring up for their chicks.

In order for the trees to bear branches low enough to suit a nesting female warbler, the jack pines must never get much larger than a big Christmas tree. In woods where the pines are even a bit taller, or among trees other than jack pines, or even among perfectly suitable jack pines growing in a stand that does not suit the warbler's desire for a patchwork of wooded areas and clearings, the finicky warbler simply will not nest.

Wildlife biologists have known that much for decades. But they had long been puzzled about precisely why the warblers would not even extend their ranges to stands of jack pines elsewhere—say, in Michigan's Upper Peninsula—even when conditions seemed perfect. In the 1980s, Cleland determined that, right here, at this sandy spot at the middle knuckle of the Michigan mitt, climate and soil interact to produce a woods perfectly suited for Kirtland's warblers. In short: these woods historically have burned down more frequently than even other fire-adapted jack-pine forests.

"The burning cycle here was a ten-thousand-acre fire every twenty-eight years before Europeans settled here," Cleland says. "The warblers move into a stand when the trees are around age eight and move out around age twenty-six." Jack pine, an unusually hardy species of tree, typically grows in climates where the days between freezing weather are few. But ironically, the jack-pine-dependent warbler, he says, will assiduously avoid nesting where frost is likely during the nesting season. So it avoids low swales that can become frost pockets.

"The growing season here is already short," he says, "only about eighty-five days. But for jack pine, it's still warmer than the warblers will find in Minnesota, or Wisconsin, or the Upper Peninsula."

So part of the answer is that it has historically been too cold north of here for the warblers to breed successfully. Meanwhile, the climate east and west and even directly north of this spot is moderated by Lakes Michigan and Huron, which join at the top of the mitt at the Straits of Mackinac. Even a few dozen miles west of here, for instance, the frost-free season nearly *doubles* thanks to the climate moderation from Lake Michigan. Soils are far richer there, too. As a result, the jack pine cannot compete with other tree species that proliferate in these less harsh environments, and these more temperate areas are virtually devoid of the trees. (In fact, while the winter climate on this outwash plain can be brutal, the moderation from the Great Lakes is so dramatic only several dozen miles to the west that the area is a leading peach- and cherry-growing region.)

This particular forest burns on fast cycles even for jack pines because the searing summer heat makes this jack-pine woods even drier and more incendiary than they might be, according to Cleland. But the "Dickey bird" evolved not only to cope with such frequent, apparently devastating fires; it virtually *needs* them.

In short: without frequent and devastating fire, and subsequent regeneration of young jack pines, in just the right climate for nesting, no Kirtland's warbler. For this species to live, the forest must be virtually destroyed, and then must quickly begin to regenerate, just as it has since the glaciers retreated from this region.

The Kirtland's warbler is protected by the U.S. Endangered Species Act. That means that the Forest Service and the state of Michigan, which holds state forestlands near here, too, must go to heroic ends to try to preserve the bird. That, in fact, was the reason foresters were staging a controlled burn in 1980 in the first place: by first logging out the trees, and then burning the ground litter, they were attempting to replicate the effects of a fire without actually starting a huge conflagration that could torch nearby settlements and cottages and homes, or risk human lives.

Some lessons emerge from decades of attempts to preserve the warbler here, Ennis tells me. He readily acknowledges that within a few years of the Mack Lake fire, Kirtland's warblers began to proliferate here as they had never before. But fires like the Mack Lake inferno are simply unacceptable because of the threat to private property and human life, he insists. If the bird is to survive, wildlife managers must play God, manipulating the environment as best they can to suit the bird's needs.

They have learned, at least somewhat successfully, to do just that. On our way to this spot, Cleland and Ennis and I had flown in an old, sturdy, Forest Service fire-fighting airplane over Kirtland's warbler country. Below us, the effects of playing God were evident: arrayed across the forest's entire landscape, on both federal and state forestlands, were carefully arranged designs—blocks or undulating waves of young, growing jack pines, interspersed with open clearings. As Cleland explained it, foresters were manipulating these woods to try to mimic the patterns in which the pines would naturally tend to regenerate after sweeping fires.

In their manipulations, foresters simply sell off swaths of trees to loggers as the jack pines reach about thirty years. (Perhaps surprisingly, in light of the long lives of the Northwest old-growth trees, jack pines are moving to the ends of their lives already by then. In a natural stand, they would soon be weak enough to be unable to ward off pests, like the jack-pine budworm, which can proliferate to such an extent in an old and weakened jack-pine woods that I have been able to stand silent in a grove and listen to the hiss of millions of bits of feces—frass—from the budworm larvae raining down.)

Next, loggers clear-cut the parcels they had bought and haul away the boles, to be made into, say, paper pulp. With that fuel gone, foresters then conduct their controlled burns. If jack pines lived for centuries on a landscape free of fire, their needles would pile up several feet deep on the forest floor, for they are resistant to decay. But the controlled burn, like a naturally caused fire, releases their nutrients back to the soil. After the logging, and the burn, foresters plant new seedlings on the charred ground in a pattern presumed to suit the warblers' nesting and foraging needs.

Yet, while the wildlife managers and foresters have worked diligently to try to mimic nature, Kirtland's warbler reproduction was marginal at best in the re-created landscape—until the Mack Lake fire. By 1987, seven years after the conflagration, wildlife biologists were able to count, by listening for their territorial songs, only 167 males. But by then, the trees in the Mack Lake burn were just about to reach the proper age for warbler nesting. And as the trees in the Mack Lake burn site reached the proper age, the warbler population began to soar. By 1995, the population had more than quadrupled, to 759 singing males. At least half of the birds were occupying nests within the site of the Mack Lake fire.

Birds *were* breeding in the human-designed landscapes as well—but not as effectively. Still, given the virtual impossibility of purposely staging huge fires on a landscape interwoven with human settlements, the logging and prescribed burn program seem to be the most workable alternative.

Clearly, one of the reasons Forest Service officials had been so eager for me to see this site is that it offers a neat package, with the tiny warbler serving almost as a public relations representative for vigorous logging. Under the current management plan, an endangered bird gets its mandatory habitat, and local loggers and wood-products companies get the wood fiber. And state representatives of both the Audubon Society and the Sierra Club, environmental groups that have ardently opposed clear-cutting elsewhere, acknowledge that they approve of the plan as the best compromise between the ecosystem's needs and pure practicality.

But in many ways, the jack-pine forest is unusual. Even as a natural landscape, its ecological behavior certainly cannot be compared directly with a mixed and diverse old-growth stand. Its natural pattern of succession is not to move toward great diversity and complexity. When it burns, it burns quite thoroughly. In nature, a new stand that grows on a burned site is virtually as even-aged as an industrial plantation.

But scientist Cleland insists that the story of the jack-pine woods and the warbler symbolizes something more sweeping and important. He says he shares one conviction with the Andrews scientists, as well as a host of other ecologists: that by truly understanding how a forest ecosystem works, and particularly by understanding its natural patterns of disturbance, one can design forestry that works in harmony with natural patterns and processes.

Cleland insists that by following ecological principles more closely, "there are ways to cut down trees without harming a damn thing."

Cleland says that does not mean he is advocating clear-cutting everywhere, or even widespread use of the kind of intensive, industrial forestry his employer has sometimes infamously practiced in the past (although he suggests that it might make sense to use as intensive, industrial-style "fiber farms" a portion of public forests that provide little value in terms of, say, biological diversity, in order to protect other blocks of woods more completely, and still produce wood fiber). He says that following nature's pattern would mean, in many forest types, avoiding clear-cutting like a dread disease. By way of example, Cleland and Forest Service ecologist Jim Jordan would later lead me into a stand of northern hardwoods in Michigan's Upper Peninsula.

Perhaps one hundred years old, and a wholly different kind of ecosystem than the jack pine, that stand represents what Cleland calls the asbestos forest. These northern hardwoods—typified by trees like sugar maple, yellow birch, and eastern hemlock (which is not a hardwood, after all), did not evolve at all as ecosystems driven by large, sweeping fires. Underlain by moist soils, these ecosystems are simply not susceptible to fire, even in dry years. Nor, for that matter, does "natural disturbance" in a system like this even remotely resemble the sweeping destruction so common in jack-pine stands. In this ecosystem, Cleland says, disturbances might be fairly frequent, but they are often so minimal as to nearly escape notice. In the hardwoods, disturbances come suddenly from windstorms, or localized insect or disease infestations, or combinations of several factors.

The scale of the disturbance: "Here or there, a single tree or a small group of trees fall over," he says. "That's about it." But in time, the patchy, small disturbances are just the ecological ticket for the forest to continually renew itself. Where single trees, or small groups of trees, fall to earth, sunlight enters the forest, and young trees can began to grow in the gaps, maintaining a diversity of tree sizes throughout a natural northern hardwood system. Meanwhile, young shade-tolerant species, most notably the sugar maple, can thrive in the darker subcanopy, slowly working their way to the light.

If one wishes to remove timber from such an ecosystem, Cleland advocates again following nature's own pattern. In fact, ecologist Jordan had gone a bit beyond that here. Because loggers in the nineteenth and early twentieth century axed these forests from horizon to horizon, almost no old growth remains in the northern hardwood region, which includes the northern Great Lakes area and parts of upstate New York, New England, and eastern Canada. One of the few remaining old-growth parcels, in fact, lies near here in a protected national forest wilderness preserve known as the Sylvania Tract. Mindful of that, and familiar with the old growth's natural structure from his own rambles in the Sylvania, Jordan had designed a logging program for this stand that attempted to mimic old growth. The plan had directed loggers to remove some trees, to fell others and leave the logs on the forest floor, and to girdle (remove a ring of bark from) others and let them die to become snags where they stood.

"You can imagine the reaction I got when I told them they had to cut down valuable trees or kill them and leave them behind," he declared.

He had, in fact, also insisted that some of the largest and most vigorously growing green trees be left behind. In essence, the woods had been merely lightly thinned. What remained on the site was a logged forest that simply did not look like one. In fact, it was lovely to the eye. Since it had been selectively cut, with perhaps only 10 percent of the trees removed, it was nearly impossible to tell that just two years earlier logging had occurred here. The few trees that had been cut down had opened up gaps in the canopy, where sun now lanced through. In those spots, young trees were already growing. Indeed, although the scale of the trees was far smaller, this woods already had some of the *feel* of Andrews-like old growth—and it clearly would have even more of that feel as those young trees grew higher to form a multistoried canopy.

Jordan, however, emphasized that "this is *not* old growth. We can't create old growth, only time can do that. But we're trying to prove

that we know enough to at least try to create some of the attributes."
(Jordan, in fact, made that point to me again later, to be sure I got it:
he was not trying to suggest that a forester could create old growth.
Only time can do that. Nevertheless, a few months later a senior
Forest Service official at the eastern regional office in Milwaukee
would suggest that Jordan's experiment proved that old growth
could indeed be created.)

But although these woods were hardly the real thing, Jordan and
Cleland believe that they will prove to be an invaluable stopgap to
provide habitat for species that depend on old-growth structures—
like forest songbirds that depend on niches in, say, various reaches of
a multistoried canopy. By way of contract, and directly across the
road, we looked at another stand of northern hardwoods. Virtually
the same age, it had not yet been logged in any way. It was an even-
aged forest—an uncharacteristic condition for these hardwoods, with
a dark forest floor, no gaps in the interdigitated canopy, and virtually
no hope of much diversity in structure in the short term. (In the long
term, the trees in this woods, spaced too close together, would sort
themselves into a more diverse stand through a natural process of
thinning if no one logged the stand. In a fierce battle for sun and
water and nutrients, some would become stressed and die, unable to
ward off insects or disease. The survivors could flourish without such
intense competition, and younger trees could grow in the sunlit gaps
left by the fallen. Over decades, in other words, this stand too could
become more diverse, and move toward becoming old growth itself.
Jordan had simply attempted to accelerate the process, and, of course,
sell some valuable timber in the meantime.)

Although many environmentalists are skeptical that forestry, as a
science, can ever accomplish much more than mowing down wild
forests and, at best, converting them to ecologically simplified tree
farms, ecologists like Cleland and Jordan are part of a growing shift
in perspective among scientists both within and without the Forest
Service.

Although the deeply entrenched assumptions of turn-of-the-
century "scientific" forestry still dominate forest management, signs

and signals are everywhere that a new generation of thinkers has already begun to change that paradigm. This new breed of forest scientist is shifting the schema away from a mechanical, industrial approach that looks at forests like an efficient farm, and toward something that might even be called an ecoforestry.

Jerry Franklin, and the Andrews group, even dared call their groundbreaking approach the New Forestry.

The conception of a New Forestry had everything to do with another, far more dramatic forest catastrophe in 1980, the same year as the Mack Lake fire.

At that point the Andrews team had been working together for a decade. Some had left the team. But, remarkably, many of the core group of about twenty had stayed. The team had begun to acquire some semblance of a more permanent home—a collection of used, ramshackle trailers, and a captive campground dubbed "the gypsy camp," near one boundary of the experimental forest.

Franklin had spent much of the 1970s focusing his energies on team building, even morale building, encouraging researchers from disparate disciplines to talk and work together. He organized overnight campouts of team members, with activities that ranged from Frisbee tournaments to bull sessions around the campfire whenever it seemed like a good time to try to synthesize research.

The International Biological Programme funding had expired in 1976, but by then the group's research had caught the attention of the National Science Foundation, which had begun providing its own core of funding for the project. In fact, the research adventure in the Andrews would become one of the key models for what is now a nationwide network of similar sites, ranging from a forest in tropical Puerto Rico to a desert in New Mexico to a tundra ecosystem in Alaska, called the Long-Term Ecological Research Program.

But resources remained scarce, and the condition of facilities and, often, equipment, says Swanson, was often almost absurdly inadequate. Even with help from the National Science Foundation, researchers were constantly scrambling for supplies and funds for research assistants or computer time or laboratory analyses of samples.

Denison's miner's lamp cum acetylene generator might have served as a fitting symbol of the spirit of innovation on the cheap. The most lasting symbol of the limitations of resources came one day when a graduate student research assistant stepped into the shower in one of the beat-up trailers and promptly found himself, along with the rotted floor of the shower, plunging to the ground.

Swanson thinks the sense of team spirit was actually helped along by the sense that the team was conducting cutting-edge science in the face of adversity and, as far as grant funding went, near impoverishment. "There was a spirit of doing it in spite of the system. You figured you were out of the mainstream, that the system wasn't going to give you much support, and that was energizing."

Early in 1977, Franklin brought the Andrews scientists together for what seemed like a limited and straightforward project, but one that would ignite a slow, long fuse that would eventually make the preservation of the ancient woods a central environmental cause. Art McKee recalls that Jerry Patchen, a senior forester at the Siuslaw National Forest, which lies in the Oregon Coast Range, about eighty miles west of the Andrews, was confronted with an intractable problem. New Forest Service policies mandated that forest supervisors endeavor to protect at least some old growth, if it existed. McKee says that definitions of old growth were so sketchy that they were mainly along the lines of "I know it when I see it."

Patchen said (as McKee recalls), "How can I manage old growth when no one can tell me what, exactly, old growth is?"

Franklin promised to try to provide an answer—a formal definition of old growth based on a synthesis of what the disparate researchers had found so far. The group gathered around a table in the two-story headquarters building at the Wind River Experimental Forest in Washington, not a mile from the future site of the canopy crane, and began to piece together their description of just what old growth really is, and how it differs from a younger forest. Over the next several months Franklin pieced together bits of text and comments from an array of team members, and eventually assembled a monograph called "The Ecological Characteristics of Old-Growth Douglas-Fir Forests."

When he had first proposed the Andrews whole-ecosystem study, Franklin had endured resistance to the entire idea because foresters nearly universally assumed that all the old growth would be gone in a matter of a few decades anyway, and, since it was an inefficient and decadent forest, gone for good reason. But in the new paper, the Andrews researchers clearly were stating—even if in the muted tones of scientific prose—that a great array of structures and ecological processes in the old growth had great biological value. Moreover, the researchers implied that the loss of old growth had perhaps already become extreme.

Obviously, important characteristics of old growth were old trees and big trees. As a rule, a site with an abundance of trees two hundred years old or older would have attained formal old-growth status, the team decided. (However, the report noted that an old-growth forest was not necessarily a forest in the final, or climax, stage of succession. In the Northwest, the huge, ancient Douglas firs that are such a common component of old growth are actually subclimax trees. Absent any kind of disturbance in a stand, they would be gradually replaced, after centuries, with climax species such as western hemlock. However, in most cases, forests in the region never reach the climax stage because fires and, to a lesser degree, windstorms tend to drive most stands back to a subclimax condition.)

The report focused on other key components. The old-growth forest was structurally diverse—an array of not only old, giant trees, but layers of younger trees and shrubs growing in the subcanopy—in the sun-flooded openings where old trees had fallen. It was a system also defined by big standing snags that remained in place for astonishing lengths of time (team members had determined that western red cedar snags could stand for more than a century, and Douglas fir snags for up to seventy-five years). And most important, it was a system driven by its supply of logs—coarse woody debris on land and, especially, in streams.

The report's implications were clear. The survival status of species in America had become an issue of increasing public concern, a wave that crested in the early 1970s with the passage by Congress of the Endangered Species Act. Now, said the study, "there is considerable logic in maintaining entire stands or small drainages for old-growth attributes. The old-growth ecosystem is a system of many interlinked components, including organisms ... some organisms or functions may depend on an intact old-growth forest for their perpetuation." Furthermore, the Andrews scientists reported that remaining old growth not protected in parks or other reserves, such as designated wilderness zones, was already well on its way to vanishing. Existing reserves, they wrote, "occupy less than 5 percent of the original land-scape, and the end of the unreserved old-growth is in sight."

The document was put together principally by Franklin, soils ecologist Kermit Cromack, Bill Denison, Art McKee, Chris Maser, aquatic biologist Jim Sedell, Fred Swanson, and Glenn Juday, a doc-toral candidate at the time of the 1977 meeting.

It would be 1981 before the report was formally published by the Forest Service research branch. Perhaps some of the delay could be excused as the usual ponderous behavior of a huge bureaucracy. But some critics believe that Franklin and perhaps others in the group were well aware that the report's fundamental findings had grave implications for forestry itself in a region that produced more than half of the nation's lumber and wood fiber supply. The report, after all, shattered the myth of the dying, unproductive "cellulose ceme-tery." And it thus implied that much of conventional thinking—the very thinking that was driving the conversion of natural woods to plantations—had deep and fundamental flaws.

Yet word of the Andrews team's discoveries began to spread through the environmental community well before the monograph was formally published. In 1978 Glenn Juday read a paper summa-rizing many of the team's findings about the value of old growth at an environmental conference at Lewis and Clark College in Portland. A student at the conference, one Cameron LaFollete, an undergraduate at Reed, wrote a paper summarizing Juday's presen-tation for the Oregon Student Public Interest Research Group. That report began to circulate among environmentalists. The term *old growth* had not been in common use much outside the scientific com-munity. But many environmentalists already shared a growing loathing of the clear-cuts that seemed to expand by the year in a mountainous region where the scars of logging were all too evident from almost any vantage point. Now, it seemed, science was accumulating data to support that visceral reaction.

Then in 1980, Chris Maser and Jim Trappe, the two scientists who had pieced together the commensal relationships between tiny rodents, and truffles, and old-growth trees, published a booklet about their findings, called *The Seen and the Unseen World of the Fallen Tree*. Beyond detailing the intricate and interdependent ecology of these species, they were among the first federal scientists to fire a warning that forestry as practiced in the region was fundamentally flawed.

"What will happen to the Douglas-fir ecosystem when fallen trees are no longer added, as will be the case under intensive forest management with increased utilization of wood fiber? And what will happen under short rotation management, when large trees are no longer produced?

"We must not sacrifice the options of future generations on the altar of cost-effectiveness," the scientists added.

On May 18, 1980, came the event that ultimately would alter the Andrews team's view of forest succession, and even their view of the practice of forestry, more profoundly than their accumulating knowledge of the old growth. It was a disturbance whose main force visited an expanse of northwestern forest and meadow over a period of only a few hours, and yet caused almost unimaginable devastation. The research that followed it would lead quickly from discovery to recovery, and to a series of dramatic new insights about how ecosystems are reborn.

It was a clear, sunny, azure-sky morning in the Western Cascades when geologist David A. Johnston flipped on his two-way radio to report the results of that morning's observations of Mount St. Helens, about fifty miles northeast of Portland. The measurements he had taken that morning showed nothing particularly unusual about the mountain—or at least nothing he and other scientists didn't already know. The volcanic mountain had become active almost exactly two months earlier, after a long, 123-year dormant period. On March 16, a series of tiny earthquakes, and then a somewhat larger one—about

4.2 on the Richter scale—had shaken the mountain. In the next ten days the mountain had shuddered repeatedly, with hundreds of small to middling earthquakes shaking the area, and on March 27 the volcano erupted, throwing a steady stream of volcanic ash six thousand feet into the air.

Johnston and other volcanologists believed that a series of rhythmic, long-lasting tremors that began about March 31 were signs that gases or magma or both were moving powerfully beneath the mountain, and probably signaled a far more substantial eruption. And yet the volcano was almost coy. Late in April, the existing eruption shut down. On May 7, it began again, then dwindled again to nothing by May 16. Still, the long tremors were continuing. Seismographs had recorded nearly ten thousand single earthquake jolts. Meanwhile, it seemed clear that magma was accumulating under the mountain's north face, for that slope was visibly bulging.

Johnston's major task that morning was to focus a laser beam on reflectors geologists had scattered around the bulge. Once again, his measurements confirmed that the bulge was growing.

Johnston finished radioing in his results at about seven that morning. Near eight-thirty yet another earthquake—this one larger than the others—began to shake the mountain. In a sort of slow motion, a giant block broke off the north side, then another. Then the entire bulging north flank of the mountain collapsed, and the largest landslide recorded in history began to roar down the mountainside at speeds of more than 150 miles per hour. It was the sheer mass and momentum of the slide that carried it over the 1,150-foot ridge that Fred Swanson would ten days later regard with pure awe.

Inside the volcano, enormous pressures had been building up for weeks. Now it was as if someone had punctured an overinflated and immense balloon. At twenty seconds past 8:32 A.M., the mountain face exploded in a debris-filled sideways blast that thundered outward. Calculations would later show that the blast reached speeds just under the speed of sound. Within about eight miles of its center, the blast leveled virtually everything it encountered. As far as nineteen miles away, it caused at least spotty damage: even at that outmost reach, trees were singed and killed where they stood by superheated gases. Johnston's observation post was fully six miles from the volcano. But he would not survive the lateral eruption's ferocity.

Moments later, the volcano erupted vertically and a column of ash rose twelve miles into the sky, forming a giant mushroom-shaped cloud. Lightning lanced out of the cloud and promptly ignited dozens of forest fires as the cloud began moving east and north.

Before noon, the spreading cloud of ash had turned midday to twilight in Spokane, some three hundred miles to the northeast, where automatic streetlights flicked on, then remained on all day. Near Ritzville in nearby eastern Washington, fully two inches of ash rained to the ground. Two days later, fine bits of Mount St. Helens ash settled on the east coast of the United States. And within two weeks, geologists in the state of Washington were able to collect ash that had been thrown high in the jet stream and had, by then, worked its way completely around the planet.

But the vertical eruption was not the last of it. Hot pyroclastic flows pelted down the north slope, glowing blends of gases, magma, and rock that approached temperatures of about 800°F. Moving at hundreds of miles an hour, the pyroclastic flows melted the mountaintop snow and glacial ice in a flash, and, suddenly, immense mudflows made of meltwater and glacial debris began pushing their way down part of the mountainside, ultimately clogging the Cowlitz River below.

Fred Swanson reported his wonderment to Franklin after returning from his expedition to the scarred mountain slopes with the early geologic team. And Franklin quickly realized that as much as any event in history, the eruption of Mount St. Helens could provide the team a chance to examine a core ecological process in action. But unlike Henry Cowles's adventure on the dunes, reconstructing the successional process by inference here would not be necessary. If the team moved quickly, it could survey the site before colonization by new life from outside the blast zone began—or so Franklin thought—and then follow the process over months, even years.

Franklin put it this way: "What an opportunity, I thought. At last, we would be able to study ecological succession—the changes in plant and animal populations that occur over time—starting from a totally barren area. Over the years, we could trace how long it took various organisms to reenter the area and in what order they would appear."

Within days a core group from the Andrews team helicoptered into a site high on the volcano that should have been as barren as any place could be. It was, indeed, an ethereal sight. A world where trees had towered high now was a land of only jagged, giant stumps, a horrible otherworld painted with a uniform gray-white layer of volcanic ash, called tephra. Art McKee recalls his first look at it all, from a helicopter that shot first across an expanse of green forest that had barely been dusted by ash. The researchers he was with were all chattering to each other over their headsets. Then the helicopter rose over a ridge, and suddenly the devastated landscape lay gray-white, a pallor of death, barren and eerie. The scientists went dead silent.

But even as they stepped out of a cloud of ash from the rotating helicopter blades, the scientists spotted tiny new sprouts of fireweed poking their way through the ash. Within hours, Andrews team members found living ants, beetles, and salamanders under patches of not-melted snowpack, even a few seedlings and saplings that had somehow survived.

"The reinvasion of organisms from undamaged nearby areas would certainly be important [to ecosystem recovery]," Franklin later wrote. But what astonished him most was that "for the most part, the organisms were already there." Even before the site was colonized by species from outside, in other words, its ecological recovery had begun.

Recovery would proceed much more slowly at some sites than others, however. Some landscapes were buried under a thick blanket of mud and debris, others merely with volcanic ash. Some of the most devastated and bleak had been blasted with the near-700°F pyroclastic flows. But it seemed remarkable, nevertheless, to find plants poking up and insects crawling even in the blast zone. (As an indication of the fury of the blast, McKee offers this: during their time on the volcano, the researchers found an immense tree, upside down, roots pointing skyward, its top stuck in the mud in the middle of small Grizzly Lake. It was an old-growth Douglas fir. But no old Douglas firs had grown within miles of Grizzly Lake. Apparently the immense tree had been blown through the air at least seven miles.)

The first trip to the mountain turned into several more—a long "research pulse," the team dubbed it—for Franklin and the Andrews scientists realized they were finding startling linkages with their discoveries in the Andrews forest. One sterling example: gophers are generally considered to be parts of an ecosystem that can retard plant succession, for they are industrious diggers and devourers of the vital roots and other underground parts of plants. But the pocket gophers that had survived in the blast zone by remaining burrowed under the ground turned out also to be industrious tillers of soil, blending the sterile volcanic ash with the fertile soil below. The blended soil had a far greater ability to hold water, and held far greater levels of critical nutrients like phosphorus, nitrogen, and carbon, than the raw tephra alone. Perhaps more telling, the gophers turned out to be consumers of the mycorrhizal fungi that had already begun to spread their hyphae rapidly through the soil, including the very burn-site fungi Swanson had seen in the small hole he'd dug on his first trip to the volcano. Through their fecal pellets, the gophers were serving as major dispersers of spores. The gophers, it seemed, not only were not retarding succession, they were driving it forward.

A host of different kinds of organisms had survived the horrific eruption. Some small mammals, such as the gophers, had survived simply because their underground burrows served as bomb shelters. Some seeds had apparently survived because they had long been buried in the burrows and tunnels of the small mammals. Plants like blackberry that sprout from rhizomes, which are rootlike lateral stems, regenerated by simply pushing new sprouts up through as much as a foot of ash. In some places, where the tephra or landslide debris was thicker, erosion actually helped the regeneration process along, washing away loose ash and debris and carving gullies down to the original soil, where new plants could take root. In places where the snowpack remained unmelted, the snow actually protected seedlings of trees below, as well as shrubs and small plants and at least one type of mammal, the tiny deer mouse. In fact, entire plant communities of some high subalpine meadows, covered by snow and a coating of ash, pushed through the ash after the snow melted and burgeoned to life in the mountain's next brief summer, showing little effect of the blast. A host of amphibians—frogs, salamanders—and some aquatic invertebrates, such as crayfish, overwintered and survived deep in the mud of lakes and streams. And once again, downed logs played a vital role: they provided the usual damp refuge for a host of fungi and arthropods.

In the blast zone, nearly everything alive aboveground had been killed. But even as the organisms that had survived emerged and began to reestablish themselves, life also began to move onto the devastated slopes: a rain of spores and seeds of plants and trees blew in on the winds; as vegetation sprang back to life in the meadows, birds returned. And even herds of huge Roosevelt elk began moving across the ash-covered slopes.

Previous theory about the pattern of succession on supposedly barren ground had emphasized just this sort of invasion and colonization of such plants and animals from outside. But the scientists were struck by the fact that residues of the life that once thrived here had survived, even in some of the most devastated areas. And it was these residues that seemed to be key to early stages of the wrecked ecosystem's recovery.

"Early patterns and rates of recovery at Mount St. Helens are strongly related to the abundance of surviving organisms," Franklin would write later, adding that "residual organisms and dead organic matter need more attention in successional theory."

It would, in fact, be the members of the Andrews research team themselves who would soon pay the most attention of all to these residual organisms. They would give them a name: biological legacies. And the team members would come to suggest that these legacies were key to a new version of forestry, based on nature's own patterns of disturbance and recovery.

9

IN THE 1970s, and into the 1980s, as the Andrews scientists were refining their studies of the ancient woods, a new crosscurrent was washing its way into the venerable science of ecology. With wild habitats falling to the chain saw or the plow or the bulldozer seemingly everywhere, concerns had been growing for years about the plight of endangered species. The Endangered Species Act, passed in 1973, offered at least some hope of systematic protection in the United States. A key international treaty, called the Convention on Trade in Endangered Species (CITES), offered some hope that it would control trafficking in species or their coveted parts (elephant ivory, rhinoceros horns) internationally.

At least ecologists thought they understood much of what was causing wildlife populations to decline and, in some cases, to slide into extinction. Despite the attention generated by high-profile stories of poaching and overhunting, it seemed clear that habitat destruction accounted for almost all of the problem.

But by the late 1970s, an entirely new subdiscipline of ecology was on the rise—one that suggested that issues involving species' extinctions and the overall loss of the planet's biological diversity were more complex than could be explained by habitat loss alone. The discipline claimed the name conservation biology. Some of this discipline's founders and proponents described it as the biology of scarcity. Focused more narrowly than the broad scope of ecology, conservation biology focused on issues relating specifically to dwindling biodiversity, that is, vanishing species or, on a finer scale, vanishing genes within species or, on a larger scale, vanishing natural assemblages of species in communities. It was a science that not only analyzed why biodiversity might decline, but attempted to develop models and schemes to halt the slide.

Conservation biology had its roots in landmark research by scientists Edward O. Wilson and Robert MacArthur. Ever since Darwin's visit to the Galápagos, the biology of species on islands has been a key area of fascination and focus. In the 1960s, Wilson and MacArthur began to look at the survival of species on islands and offered a groundbreaking analysis in their 1967 book, *The Theory of Island Biogeography*. In essence, the two biologists had found that when species are restricted to islands, the sheer numbers of species that can survive are far more limited than simple calculations about their habitat requirements would suggest. Key research focused on the myriad islands of the New Guinea shelf, for instance, which had once been part of the southeast Asian mainland. The fossil records showed that the islands had once been inhabited by a similar diversity of species. But, over thousands of years, all had lost species. Diversity had declined, in fact, in direct relation to size: each tenfold decrease in island area led to an approximate halving of the number of species that could survive there.

By the mid-1970s, ecologists working with terrestrial systems began to suggest that wild reserves—national parks, refuges, forests—had become de facto islands as well, islands of wild habitat surrounded by the artifacts of civilization: farms, ranches, cities, roads. The reasons that species on these "islands" were at risk varied. Some of them had to do with the complexities of population ecology. First, species restricted to small habitats necessarily must restrict their breeding to fewer potential partners, raising the specter of physical defects and deformities (including infertility) from inbreeding.

Despite widespread knowledge about problems in humans from close inbreeding (for instance, the well-known problem of hemophilia among some branches of European royalty), science was far less clear about the effects of inbreeding in small populations of animals. But working with the National Zoo's unusually detailed pedigree records for more than forty species from nearly as many genera at the National Zoo, Smithsonian Institution geneticists Jonathan Ballou and Katherine Ralls compared survival rates among animals with related parents against those with unrelated parents.

They later wrote, "The results were dramatic."

For instance, they found that among ring-tailed lemurs (small primates from Madagascar), more than 30 percent of inbred offspring died within six months of birth, compared to fewer than 20 percent from the noninbred births. Among black spider monkeys, the numbers were even more striking: nearly 60 percent of the inbred infants died, compared to only about 18 percent of those not inbred. And among members of a population of a hoofed mammal called the scimitar-horned oryx, the numbers were most dramatic of all. Although only about 5 percent of the noninbred young died in the first year, 100 percent of the inbred offspring perished. And in fact, in forty-one of the forty-four species, the inbred offspring died as youngsters at significantly higher rates.

But conservation biologists would conclude that inbreeding is only the beginning of a matrix of problems for small populations in islandlike settings. Computer models projecting animal populations over time quickly began to uncover another grave possibility. Many populations fluctuate naturally. Certainly, a devastating storm, wildfire, or, for that matter, an erupting volcano can either wipe out or severely reduce an already small and limited population. Population models suggest that, over time, isolated pockets of a few members of a species can simply blink out of existence, one at a time, like burned-out Christmas tree lights, as a consequence of genetic problems or simply the pressures of random (ecologists like to use the word *stochastic*) localized declines, or a combination of the two. (Consider, for instance, what would happen if a local population of, say, red-backed voles, was reduced to a functional island with only ten breeding-age members. Perhaps a particularly devastating fire affects the habitat, with only half of the tiny population destined to survive. But, of the five that survive, only two, or one, or none might be females, absolutely reducing the effective breeding population to as little as zero. Even if a better ratio of sexes survives, this population, reduced even further, is even more intensely susceptible to inbreeding. Or yet another random minidisaster, something as common as a disease or an untimely encounter with a predator, could wipe out one or more years of breeding.) The problem was particularly acute when tiny habitat islands were so separated from each other that a local extinction meant no possibility of recolonization of even superb habitat.

And there are still more problems, perhaps most notably this: tiny terrestrial "islands," in highly fragmented landscapes of once-wild ecosystems, can be even more vulnerable than they appear because so much of their apparently sound habitat is "edge." The edges of forest "islands," for instance, are more susceptible to being blown down in high winds. Worse than that, the first several feet of forest edges often are ripe for invasion by certain kinds of predators and parasites from the disturbed and more open "sea" around them, including predators and parasites that could never survive in a large block of deep interior forest.

Working in the intact blocks of forests of the Great Smoky Mountains National Park and more fragmented forests in Maryland, for instance, ecologist David Wilcove in the early 1980s recorded dramatic declines in breeding success among songbirds in fragmented islandlike habitats. In studies that have been repeated with similar results in other regions, Wilcove found that a key culprit is a so-called brood-parasite bird, the brown-headed cowbird. A denizen of clearings and open fields, the cowbird lays its eggs in the nests of unsuspecting songbirds, sometimes even rolling the songbirds' own eggs from the nests. The young cowbirds, once hatched, are not recognized by the songbird mothers as interlopers. They eat voraciously and grow quickly, and songbird chicks seldom survive the experience of sharing a nest. Songbirds can suffer similar losses from predators ranging from raccoons and feral house cats (which now, incidentally, roam the American continent by the millions). Deep, mature forest tends to exclude these opportunistic nest predators as well the parasitic cowbirds, all of which are more inclined to occupy more open areas like fields, or young forests, or the margins between forest and field, or even suburban lots.

The upshot of the edge problem is that it effectively reduces the size of an already isolated "island" fragment of wild forest. A round "island" of habitat one hundred yards in diameter is, essentially, 100 percent edge if the "edge effects" extend inward as little as fifty yards. In other words, if a cowbird can penetrate fifty yards into such a block of woods, the woods is ecologically not the deep forest it appears to be, but a block of pure edge habitat. Similarly, even an extremely large oblong area of "wild" can be predominantly edge. In 1987, then-Andrews team leader Jerry Franklin and coauthor Richard Forman, a leading landscape ecologist from Harvard, calculated that if fragments of about 250 acres are denuded, cookie cutter–style, from a typical Pacific Northwest Douglas fir forest, virtually all of the interior forest would be effectively gone once only 50 percent of the forest had been felled, due to edge effects.

Hard on the heels of rising new concerns about a vanishing, or at least seriously degraded, wild, the new conservation biologists began trying to determine not only how severe problems such as reserve size, or small population hazards, or habitat fragmentation might be in some places, but what could be done to stem the rising tide of loss that their calculations were projecting.

One prominent conservation biologist, Larry Harris, of the University of Florida, found himself thinking just that as his commercial airline flight winged its way from Florida across the country to Portland, Oregon, in the early 1980s. Harris, on sabbatical, was stunned when he saw the west-side forests of the Cascades below him. For years, the U.S. Forest Service, the agency responsible for millions of acres of west-side woods, had followed a policy of limiting clear-cuts to no more than forty acres, assuming that small cuts would be more environmentally acceptable, or at least less objectionably ugly. Commercial forests, on the other hand, were more often subject to sweeping clear-cuts. But much of the public land of the West, including national forest acreage, and especially public forest-

lands administered by the Bureau of Land Management, is interspersed, checkerboard-style, with industrial forestland, with a 640-
acre (one-square-mile) public section for each 640 acres privately
owned. The pattern is the residue of free land incentives given by the
federal government to railroad companies in order to induce them to
spread their rail lines westward in the nineteenth century. From the
air, the pattern is striking. Often the industrial square miles have
been completely clear-cut; the federal square miles have meanwhile
been fragmented with small galaxies of many smaller, forty-acre
clear-cuts. The effect, after years of forest fragmentation, was one of
small islands on the public land, comprising a fragmented larger
island surrounded by a sea of recently clear-cut, or monoculture,
even-aged young Douglas firs, on the industrial land.

Harris quickly found his way to the Andrews and embarked on a
long study of the effects of the fragmentation of the ancient forests in
the region. In succeeding months, he and Andrews biologists Art
McKee and Chris Maser would begin folding a new ingredient into
the old-growth recipe: the notion that the ecological condition of an
entire landscape was at least as important as the sheer amount of
remaining old growth.

"How much old growth should be perpetuated is only part of the
question," they wrote. "Size and spacing are of equal importance."
Any reserve designed to protect a full range of species would have to
be far larger than the fragments current logging patterns were turning the landscape into. The scientists suggested that perhaps up to
three hundred to five hundred acres were required to provide for the
needs of small mammals.

And then, in a suggestion that presaged a coming war in the old-
growth woods, they suggested that islandlike reserves might have
to be as large as one thousand acres to provide for certain "predatory birds."

By then, Eric Forsman had been thinking about a certain predatory
bird for more than a decade. It all began when Forsman was on a
summer break. A university undergraduate in wildlife biology at
Oregon State, he had managed to secure a plum job as a fire lookout

for the Willamette National Forest in 1968. One day, as he walked near his lookout post, an unfamiliar hacking hoot came from the parcel of ancient forest. Forsman, puzzled, barked back in imitation. Suddenly two owls, members of a species he had never seen before, their breasts dappled with spots, flapped down from the canopy and into plain view.

That first encounter with spotted owls sent Forsman on a research path that would lead him, too, to the Andrews and would lead the owls to the epicenter of one of America's most intense environmental controversies. Fascination with the birds would turn into a career and make Forsman the world's authority on the spotted owl. It would lead him to catalog the survival needs of the predator in such detail that it would one day send shock waves through the Forest Service and the logging industry.

Forsman quickly discovered that scientific knowledge about the spotted owl was virtually nonexistent. Still an undergraduate, he began looking for the birds, hooting them down out of trees in order to locate them and their nests—a project that carried him into a master's and then a doctoral program in wildlife biology. It wasn't long before it became crystal clear that the little-known owl had one obvious habitat preference: ancient forests. And by the time the Endangered Species Act was passed in 1973, word about Forsman's findings, and the bird's restricted habitat requirements, had reached far enough that the owl was included on a long list of species that might be imperiled enough to be protected under the new law, someday.

That same year, in fact, the first proposed protection for the owl came under the state of Oregon's own version of the Endangered Species Act. There were then only about one hundred known owl nests in Oregon. So the state of Oregon's wildlife department proposed that three hundred acres of old woods be preserved around every nest. Yet, despite mounting evidence that the owl depended on old growth, the Forest Service did not sign the modest protection plan for four more years. Meanwhile, more old growth was falling to chain saws, and the owl population had begun to plummet. Ironically, if the agency had undertaken an effective protection and recov-

ery plan for the owl, then and there, an enormous amount of controversy within the agency and economic upheaval in the region's logging industry might have been avoided.

Forsman, meanwhile, found himself by 1979 ensconced in one of
the Andrews team's rickety trailers. For a full year, scrambling over
logs as he lugged a portable antenna rig, he radio-tracked eight spotted owls he had outfitted with tiny backpacks holding radio trans-
mitters. Twelve months of shadowing the birds clearly established
that, although they would sometimes use regenerating forests, the
birds clearly depended on old growth for nesting and, especially, its
rich supply of squirrels and voles and other prey.

The case of the spotted owl would, of course, explode into the
most furious of environmental controversies in U.S. history. It
became, through politicians and the news media, a sort of great
morality play: In one version, it was the rapacious timber companies
versus a spectacular and ecologically "tiny bird" (never mind that, as
birds go, the spotted owl is actually quite large). In another version, it
was a silly bird versus millions of dollars of economic value to be
"locked up" as protected old-growth timber, and the related loss of
thousands of timber jobs. (In fact, protecting old-growth forests to
protect the owl eventually did eliminate many logging jobs, probably
thousands. Environmentalists have been quick to point out that even
the industry's highest estimates of job losses pale beside the thousands
of jobs eliminated in the timber and lumber industry in the region
due to rapid mechanization. In a May 1989 analysis, the timber
industry itself acknowledged that, even before any restrictions on
logging occurred because of the owl, in the short span between 1979
and 1987 alone, the industry had lost twenty-six thousand jobs to
mechanization in the Northwest, or about 15 percent of total industry employment, in a trend that seemed certain to continue. Production, measured in billions of board feet, had meanwhile climbed from
less than fifteen billion to more than seventeen billion. Nor, for that
matter, do the "lost job" projections acknowledge that the jobs in
mills that were designed to process giant trees would have been lost
in many areas soon enough, anyway, as old growth vanished.)

As the 1980s progressed, the regional owl controversy would begin to go national. In 1985, University of Chicago conservation biologist Russell Lande, using a population viability model, began to analyze the effects of continued old-growth destruction on owls. He based his research on the Forest Service's own 1984 plan for protecting the owl. About the same time, I had been given a chance to toy with a computer program based on such a model, located at the headquarters of the International Species Inventory System in Minnesota, which kept track of breeding pedigrees among captive animals, and especially endangered captive animals. This particular population viability program allowed me to predict the future of the then-minuscule remaining wild population of the black-footed ferret, a small member of the weasel family that had been reduced to seventeen survivors in Wyoming. What was most striking about the program was that, each time I ran it, the results were somewhat different, since it was based on probabilities of the population growing, shrinking, or vanishing in any given year—fluctuations that can be quite random. But virtually every run led to extinction, although the particulars varied, year by year. In the end I instructed the computer to run the program one hundred times, projecting populations for a century, and to produce a probability map. In all but two of the one hundred runs, this virtual-reality population became extinct, usually in a few decades. In other words, there was a tiny, though extremely slim, chance, if conditions in the wild remained stable, that the population could breed successfully enough to survive another hundred years.

Lande faced a more complex sort of problem: there were more owls, but the owls' habitat would not remain stable. According to published Forest Service plans, old growth, the very habitat the spotted owls depended upon, was slated to diminish year by year. Lande's calculations, published in a leading journal, *Oecologia*, concluded that the logging plans, if followed, would mean an almost certain decline to extinction.

What would later be described as a social and environmental "train wreck" began in earnest in 1988. Environmentalists by then had formally petitioned the U.S. Fish and Wildlife Service to list the spotted owl as imperiled under the Endangered Species Act. Remarkably, despite compelling evidence to the contrary (but under intense behind-the-scenes political pressure from the timber industry and Northwest politicians), the agency declined to do so. In November 1988 a Seattle court harshly ordered the Fish and Wildlife Service to revisit the decision, suggesting it had been both "arbitrary and capricious" in its finding. By early 1989, the federal agency had relented and recommended that the owl be listed as "threatened" (meaning the species was not imminently endangered with extinction, but would soon enter that status if there were no change in population trends). Then, in March 1989, much of old-growth logging ground to a halt completely, when Federal District Court Judge William Dwyer in Seattle issued an injunction ordering that all logging stop on federal lands in spotted-owl habitat until the federal government proved that it had not, as environmentalists contended, violated laws requiring a full assessment of environmental impacts, as well as protection of at-risk species in the region's public forests.

The owl controversy would rage, and Dwyer's injunction would drag on, for years. During 1989 and 1990, Jerry Franklin would find himself part of a team of consulting scientists to a committee of biologists who had been ordered to try to determine the owls' actual survival needs. Notably, the main team included Forsman, by then the top owl scientist. Jack Ward Thomas, a colorful wildlife biologist who ran a research station in eastern Oregon and who had financially supported Forsman's studies all along, was tapped as chairman. According to some reports, Thomas and Forsman, especially, were initially suspicious of the newfangled ideas and population models that conservation biologists were beginning to employ. But as the scientists worked over a period of months, meeting repeatedly in the basement of a Portland hotel, the central tenets of conservation biology increasingly came into play. The entire team of biologists agreed that the best available science continued to point to the notion that, in order to survive, the owls would need large contiguous blocks of suitable forest.

In the end, Thomas would report that the entire team concurred that a scattering of small islands would not work. "We twisted and we turned every direction we could think of to defeat our own hypothesis, and we finally said, 'No, that's the scientifically credible way to go.' " The evidence for large reserves, in fact, was so compelling that not even Larry Irwin, a consulting scientist who represented the timber industry, objected to what amounted to a full consensus.

But the consensus, a plan for big islands of old growth, as Thomas would report in April 1990 to Northwest members of Congress, would mean a 25 percent reduction of timber output on national forestlands in the region, and an even greater reduction on Bureau of Land Management lands. Regional members of Congress, and loggers and timber executives, expressed shock and outrage. Environmentalists expressed skepticism that the plan would be adequate, not only for the owl, but for the protection of other species associated with old growth.

But, as Thomas would later suggest, the scientists, and the science, had "crossed the Rubicon." The spotted owl was formally listed as threatened two months later.

As controversy swirled around the owl issue and the protection of at least part of remaining old growth, Jerry Franklin and the rest of the Andrews team had begun thinking increasingly about silvaculture itself—about how to cut down trees on landscapes that would continue to be logged.

Especially after the research pulses on Mount St. Helens, the team had come up with a way to define what they were seeing there and to link it to such discoveries as the role of mycorrhizae in tree regeneration, the role of small rodents in the distribution of mycorrhizal spores, and the role of downed logs in a host of functions in the forest. The researchers named these elements of ecosystem recovery biological legacies.

The rising tide of worry about diminishing old growth had led to an obvious question: if conventional, industrial-style logging fragmented the woods and diminished the value of the woods—if indeed it removed the "biological legacies"—what was a sound alternative?

The Andrews team's research had made at least part of the answer clear. Conventional foresters had assumed that, since fire and other disturbances were nature's own method of renewal, a clear-cut was simply a more efficient version of a natural fire. But the Andrews group had proven that even a disturbance as catastrophic as Mount St. Helens left behind those legacies.

The Andrews scientists concluded that the kind of industrial-style logging that has prevailed in the Pacific Northwest—where sites were often stripped of all trees and then cleaned up, with limbs and slash removed or burned, and often then dosed with herbicides—decidedly did *not* mimic nature. While they agreed that here, as in Kirtland's warbler country, fires are a natural part of forest ecology (albeit fires that may occur hundreds of years, or even a millennium or more apart), they had by the late 1980s determined that even the most intense fires leave in their wake pieces—often large pieces—of the old forest.

"The old view was that whole stands were wiped out—one-hundred-thousand-acre holocaustic fires came along and destroyed everything in their path. But as we've looked at forest history, we've found that everything wasn't wiped out when disturbances swept through," says Dave Perry.

The team had decided to propose that the century-old tradition of "scientific" forestry itself needed to be changed. The team knew what the elements of a new paradigm should be. But they could not agree what to call a forestry that was ecosystem based. Franklin came up with a name straight out of Madison Avenue: he would call the approach the New Forestry.

"I knew some people would despise the term," he told me later. "Calling it the New Forestry suggested that there was something wrong with the old forestry."

But, clearly, there was something wrong. The researchers had realized by then, says Franklin, that any change in practices had to look at the forest as ecologists were beginning to look at all systems: as a sort of nested hierarchy. In other words, there were critical issues that had to be addressed within a small forested area of a few to several acres—what a forester might call a single stand of trees. But the flood of new knowledge from the realm of conservation biology

meant that ecologists needed to look increasingly beyond such localized issues and begin to link elements from a small ecosystem like a
stand to aggregations of those and other ecosystems in a larger and
more inclusive ecosystem—typically referred to as a landscape.

Franklin says, "This all came together really in a period of eighteen months or so. We realized that at the stand level you've really
got to be about promoting more complex systems, systems that are
structurally and organizationally diverse, and that by doing that you
can perpetuate organisms and processes. At the same time we knew
we had to begin looking at these larger landscape issues—the mosaic
of different conditions across a broader area. I guess it was the stimu-
lation of the spotted owl crisis that brought it together—the notion
that each of these pieces has to be addressed as part of a whole, and
that forestry really needed a new philosophy of how to operate."

Even through much of the more basic research years at the
Andrews, Franklin says, "I was locked into the traditional idea that
you really had only two choices, selective cutting or clear-cutting."

But now Franklin suggested to the others, "Maybe it's not just a
matter of trying to incrementally improve the old. Let's back off and
think about the whole array of premises we've come up with."

By 1989 Franklin was actively promoting the precepts of this New
Forestry. In a seminal article for *American Forests*, the journal of the
American Forestry Association, he wrote: "Forestry is at a cross-
roads. For decades we thought we knew all that we really needed to
know about forests. But in fact our level of knowledge is remarkably
superficial. . . . The traditional approach to the management of forestland has reflected a simplistic attitude that has homogenized these
complex systems. . . . Traditional practices often destroy many of the
linkages that occur in natural forests."

Franklin admitted that many of the ideas embodied in the New
Forestry had been either experimented with elsewhere or suggested
in the past by others. Nevertheless, it was striking that a group
so closely allied with the Forest Service would suggest such a dramatic suite of changes. Even more striking, the Andrews ecologists
were suggesting that timber itself should no longer drive forest management. Forestry itself, Franklin suggested, "needs to expand its

focus beyond wood production to the perpetuation of diverse forest ecosystems."

The New Foresters insisted that total clear-cutting was not acceptable in the Northwest old growth; since even intense fires usually leave behind at least a few living, old-growth giants, a scattering of live trees should be left on the site. In life they would harbor old-growth species ranging from mycorrhizal fungi to the panoply of predatory insects that seem to specialize on old growth. In time, as a new forest grew, they would serve as the new, more diverse and multistoried ecosystem's rooftops. After decades or centuries, they would die to become the future supply of the site's snags and logs. To keep biological legacies intact, said the Andrews scientists, foresters should also leave an abundance of snags and an abundance of downed logs or large limbs, letting sound, solid wood rot that could otherwise be used to make lumber or pulp.

The approach would mean an end to the notion that the best site for regeneration of a new forest was one that had been virtually sterilized with fire and herbicides. Logs would remain abundantly in streams, to provide habitat for fish and other aquatic organisms as well as a continuing supply of carbon and other nutrients. And logs would remain on land, as well, as refuges for keystone creatures like the red-backed vole and as controllers against erosion.

"It's hard for people stuck with this concept of tidy agricultural forestry to accept," Franklin told me not long after the *American Forests* article appeared, "but we're trying to suggest that a little bit of chaos is a wonderful thing in a forest."

As the new forest regenerated, it would grow into a far more diverse and complex woods than the cornrowlike plantations that now so dominated commercial woodlands of the West. In a relatively few decades, such a forest could become a mix of the old-standing green trees, the snags and logs, but also an array of younger trees. One advantage: research on Washington's Olympic Peninsula had shown that spotted owls would use similar multiaged stands much as they used old growth. "By adopting New Forestry practices, we may re-create spotted owl habitat in a matter of 90 years, rather than having to wait 200 to 250 years as with current practices," Franklin wrote.

Meanwhile, he insisted that forestry needed to move beyond thinking small. The core notions of island biogeography and conservation biology were to be brought into the designs of any forestry scheme. In the past, the Forest Service in particular had assumed that by logging in a patchwork of forty-acre cuts, harm to the larger ecosystem would be minimized. Franklin and the others suggested just the opposite: that cuts be placed beside each other, and that nonlogged lands remain aggregated in larger islands, and, as Larry Harris and others had suggested, that foresters establish forested corridors between these habitat islands through which species could migrate. Since woods beside creeks and streams often serve as natural and more easily traveled pathways, and particularly since expanding no-logging zones far from creek sides would help keep sediments from flowing into waters and polluting them, the banks of these waterways could provide some of the best such corridors.

In essence, what he was suggesting, Franklin says, was a "kinder and gentler forestry."

And maybe so. But a look at a New Forestry logging cut revealed it was anything but a pretty sight. In 1993, just outside the Andrews, Lynn Burditt and Art McKee showed me one such cut. The design certainly would have struck a logger, or timber executive, as ghastly: some twenty-seven mature, old-growth trees had been left on every acre. One hundred acres of such left-behind logs would be worth as much as one million dollars at 1993 prices. To anyone looking for some semblance of attractive forest, however, the New Forestry cut also approached the ghastly: Franklin had called such sites "sloppy clear-cuts." This one seemed, to my eye anyway, just plain ugly. In some ways, such sites look in their early months, and years, *worse* even than conventional clear-cuts. At the very least, the scattering of giant trees and the giant debris left on the site remain as a sort of grim reminder of what beauty had been lost.

Aesthetics wasn't the point, after all, McKee insisted. Biodiversity and reestablishing a healthy ecosystem were. And for that matter, a site after a forest fire—even a forest fire that is a harbinger of future health for the forest—is no prettier. Maybe so, but it is hard to imagine legions of environmentalists, already appalled at clear-cuts that sweep across the region, any more delighted by sloppy clear-cuts. It struck me that, at the very least, the New Foresters had a daunting job of education, or public relations, ahead of them.

Still, if the Andrews group had not proposed the New Forestry, it seems likely that the Forest Service would have had to invent it. By 1990, senior officers of the U.S. Forest Service found themselves under unrelenting pressure over logging in general, and logging in the Pacific Northwest in particular. Pressure seemed to be coming from everywhere. And the agency needed a relief valve. The regional fight against logging had begun with local groups like the Oregon Natural Resources Council, and in fact had, during much of the 1980s, been virtually ignored by most of the national environmental groups. But by the early 1990s, the issue had risen to the top of the national agenda.

The Wilderness Society commissioned forest ecologist Peter Morrison, who had worked with Swanson and other Andrews scientists on the fire history of the Andrews forest, to answer a question to which no one seemed to have an accurate answer: how much old growth was left? The Forest Service had claimed that fully 6.2 million acres of old growth remained in the Northwest, and the timber industry, at least, was continuing to tout those figures as proof that old growth abounded. But the Forest Service estimates seemed to be based on little more than conjecture. Looking at the region more closely, in fact, Morrison found out otherwise. Working with a combination of ground surveys, a computer-based "geographical information system," and, especially, aerial photographs, Morrison determined that there were only about 2.4 million acres left that would meet the now widely accepted definition of old growth developed by the Andrews group. Part of the reason for the disparity was that the Forest Service was counting as old growth nearly everything not recently logged or burned. But another aspect of the error was that the agency had not been able to keep up with what had become an astonishing pace of logging. Even as Morrison and his small team worked, on average just under 130 football fields of old growth were vanishing from Oregon and Washington every day.

Meanwhile, pressure from *within* the Forest Service had reached a boiling point. The Forest Service once had been an agency managed almost entirely by traditionally trained silvaculturists. Operating from Gifford Pinchot's perspective of trees as crops, the agency had focused long and hard on "getting the cut out."

For a middle-management forester, getting the cut out meant one could build a small empire—a large staff of permanent and seasonal workers. It was mandatory to get out the cut effectively to win promotions. Morever, through an almost perverse system of incentives and disincentives, the agency's programs were, and often still are, structured economically to encourage more and more logging. One fourth of fees from timber sales provided federal money, in lieu of property taxes, to schools and local governments in the very communities in which the federal foresters lived and raised their children. Meanwhile, timber sales provided even the best-intentioned district forest rangers and regional foresters and their staffs with vital funds they needed, and used, for building campgrounds and hiking trails and for maintaining wildlife habitat. (In a stunning irony, the majority of national forests have consistently *lost* money on timber sales. While these below-cost sales were far less common in the Pacific Northwest because of the high quality of the available old-growth timber, across the nation this peculiar twist meant that foresters were induced to sell timber at a loss to the public treasury in order to find funds for legitimate work, such as protecting endangered-species habitat. In the most egregious of such ironies, forest economist Randall O'Toole discovered that a forester in Idaho, desperate for money and complying with a plan to protect a piece of grizzly bear habitat, once provided it by selling to loggers another parcel of forest that was, in fact, prime grizzly bear habitat!)

But even as the spotted owl controversy was at its peak, and even as the Andrews team was proposing that forestry itself be reinvented, the Forest Service, long an agency of professional foresters, was finding itself with a growing staff of what came to be called its ologists—wildlife biologists, hydrologists, geologists, even archaeologists (many national forests contain significant Native American archaeological sites), as well as specialists in such fields as wilderness recreation.

Many of the ologists had been schooled in the milieu of the modern environmental movement. Moreover, even many of the newer and younger foresters had come of age professionally during or after the first Earth Day years of the early 1970s—a generation with a growing sense of dwindling ecological resources.

Tensions in the agency between the ologists and what some disparagingly call the timber beasts had been increasing for years when, in 1989, Jeff DeBonis—a young timber sale planner who worked, perhaps ironically, in the Blue River Ranger District (the very district that includes the Andrews)—decided he had had enough.

DeBonis had been a Peace Corps volunteer in Central America, and he had witnessed outright destruction of tropical rain forests firsthand. DeBonis's first assignment with the Forest Service was in Montana, at the Kootenai National Forest, and he had complained to agency foresters in the past about clear-cutting and burning—a scenario that he suggested had too much in common with the wildly decried forest destruction he had seen in the tropics.

But as he watched the old growth fall in Oregon, DeBonis became increasingly angry. In early 1989, he composed a long memo to the chief of the U.S. Forest Service, Dale Robertson. Forest Service employees had direct access to an electronic mail system known as the DG (the computer system had been designed by the Massachusetts company Data General). DeBonis sent copies over the network not only to Robertson, but to a long list of friends and allies within the agency. Within days, his memo had spread through the computer network like a forest firestorm.

The Forest Service, DeBonis suggested, had come to be seen as a public enemy by the environmental community. "We ... are perceived by the conservation community as being an advocate of the timber industry's agenda," he declared. "... I believe this charge is true. I also believe, along with many others, that this agency needs to retake the moral high ground."

Although he was formally reprimanded for using the agency computer system to distribute his diatribe, DeBonis did not stop there. Within months, he and a small group of others formed a nonprofit group, the Association of Forest Service Employees for Environmental Ethics (AFSEEE), and began publishing a tabloid-size newsletter called the *Inner Voice*. Articles from Forest Service employees poured in, calling attention to agency forestry practices that a rapidly swelling group of AFSEEE members saw as environmentally destructive.

Meanwhile, the timber industry was lobbying furiously to prevent further protection—or lockup, as the industry put it—of more old growth to protect the spotted owl. In some logging communities, fear of layoffs and even local economic collapse was palpable. In 1989 the loggers in the Olympic Peninsula timber town of Forks erected a giant cross beside Highway 101, with a mock grave piled high before it surrounded by plastic flowers. The cross wore a near-billboard-sized sign that proclaimed: "Here lie the hopes and dreams of our children." Sitting on the cross was an array of papier-mâché spotted owls. About the same time, Jim Torrence, the Northwest's recently retired regional forester, and I visited a café in a small, Cascades mountain logging village. A crudely lettered sign on the wall offered spotted owl soup.

Word of the Andrews group's New Forestry proposal had reached the top ranks of the agency. As Franklin had predicted, the term itself seemed to be too much for senior Forest Service bosses—all, after all, trained in the "old forestry"—to accept.

But logging trucks full of timber workers were rolling into protest rallies across the region, and even into Washington, D.C. Mainstream environmentalists were suing. And more radical environmentalists from the group Earth First! (the exclamation point is obligatory) had begun "direct action," chaining themselves to trees (sometimes high in the canopy), sabotaging logging equipment, forming barricades at logging roads. Its own ranks in revolt, and with environmentalists and their allies in Congress pressing it on one side, and the loggers and timber industry and their allies in Congress on the other, the management of the Forest Service clearly had to do *something*. Forest Service chief Dale Robertson ordered a reduction in clear-cutting, with more standing green trees, snags, and logs left on sites. In 1990, the agency announced it was embarking on a nationwide system of experiments to which it attached a name only a committee of bureaucrats could have come up with: New Perspectives for Managing the National Forest System.

Whether New Perspectives was really New Forestry or not, it immediately generated a cyclone of criticism from virtually all sides. It was based on incomplete science, cried much of the conventional forestry establishment. (So, in essence, is conventional forestry, retorted Franklin.)

William Atkinson, a professor of forest engineering at Oregon State, and one of the New Foresters' sharpest critics, called it "politically correct forestry."

"Is this going to make forests more sustainable—or is it just setting it up for some major fires and disease?" he asked me in 1990. "One thing is certain, it's going to cost a lot more."

Atkinson noted that he did not disagree with the notion that foresters should take a broader view of landscapes. "I think that concept's accepted by everybody out here at this point. I think even the private timber companies are managing on a more landscape basis. But what the Forest Service is doing is saying we want to have wildlife habitat for every critter. But that's not going to grow trees very intensively. You understand they're talking about leaving logs out in the woods to rot?

"This is a major change of goals for the national forests," he added bitterly, "a movement away from a very heavy reliance on the production of timber to a lot more concerns about habitat for animals, and much more concern about recreational use and views. That means we're not going to be producing very much timber. We're going to become a minor player in terms of the wood supply in the United States. And that's going to make a difference in terms of wood supply, and cost of wood nationally."

That July, Atkinson confronted Franklin directly in the Andrews, during a tour the Andrews group was giving for foresters and other interested parties, including Atkinson and the news media.

"New Forestry is being sold to the American people like a bar of soap," Atkinson charged. "We're getting New Forestry by decree, by dogma. We don't have Chairman Mao, we've got Jerry."

Nor did New Forestry escape criticism from the other side of the debate. "The Forest Service Fakes Reform," shouted a headline in *Sierra* magazine.

In fact, the article's author, Seth Zuckerman, had plenty of reason to be skeptical. Early on, Forest Service officials pointed Zuckerman, and other writers, including me, to a showpiece of its New Perspectives effort, a project called Shasta Costa in the Siskiyou National Forest in southwestern Oregon, the very forest where Andrews scientist Dave Perry had been trying to determine why trees in clearcuts were having so much trouble regenerating.

There, the Forest Service attempted a sort of walk-along-with-me consensus project, inviting environmentalists and industrial foresters to sit down and come to a New Perspectives–based consensus on how to log the twenty-three-thousand-acre Shasta Costa watershed, which had until then been mostly wild, mostly devoid of roads.

Local environmentalists decided to cooperate fully in the planning process for the project, if not totally enthusiastically. "We figured we'd better come up with a method that didn't trash the place," said Jim Britell, of the local Kalmiopsis chapter of the National Audubon Society.

From the negotiations proceeded a program for logging in Shasta Costa that, indeed, followed many of the prescriptions of New Forestry and indeed promised to be at least somewhat kinder and gentler. Shasta Costa was positioned between two important wilderness areas. Following the tenets of conservation biology, the plan provided for corridors for deep-forest species to move between the two wild islands. The plan aimed to remove mostly younger trees; in cases where old growth was felled, the Shasta Costa plan acknowledged the need to leave ancient live trees, as well as logs and snags.

In retrospect, it is something worse than an irony that the Forest Service so quickly decided to tout Shasta Costa as a model for its New Perspectives version of forestry. It spent astonishing amounts of public money designing and even promoting the project: about half a million dollars. It dispatched to Oregon freelance writer Janine Benyus, a former Forest Service employee who had recently authored a distinguished field guide to northwoods ecosystems, to prepare an glossy booklet about Shasta Costa, the new national model for New Perspectives. The booklet, replete with lavish color photographs, looked like a high-class corporate report, and it trumpeted the project as one that would focus on "keeping all the cogs and wheels. . . ." Shasta Costa, it suggested, would include "the full spectrum of New Perspectives techniques," including "leaving biological legacies" and meanwhile leaving in place corridors and connections "so that wildlife and plants could easily move in, out, and through the area." In particular, the glossy booklet touted an open public process that led to the design.

Still, for a time, local environmentalists seemed more convinced than ever that this New Perspective on forestry had something to offer. The Kalmiopsis chapter's newsletter suggested, near the end of the planning process in 1990, that "if logging has to be done in a roadless area, this is how we would like to see it done. Shasta Costa gave us a model."

But suddenly the project began to unravel. Forest Service employees who were key to the planning process were engaged in a whirling revolving door of transfers in and out of the district. In 1990 a newly arrived forest supervisor named Mike Lunn, the third to oversee the project, startled the environmentalists. His foresters had decided that too many of the trees in proposed logging areas were too small to generate much income. To make up for the shortfall, he proposed upping logging by 20 percent and doubling the miles of logging roads from two and a half to five. Environmentalists said they were frustrated because they could get no clear picture of just how many trees might be left behind on the logged sites. Meanwhile, the environmentalists realized that a longer-term planning document for the forest mandated that an extra eighteen million board feet of timber would be removed by 1998—this in just the few short years after the New Perspectives program removed fourteen million.

Zuckerman fairly sneered that "New Forestry and the rest of the New Perspectives program ... will likely be carried out with New Chainsaws, allowing loggers to buy New Pickup Trucks."

Soon, even as timber industrialists and allies like Atkinson were worrying out loud that these new ideas were nothing more than a sop for reducing logging in the region, leaders among Pacific Northwest environmentalists were openly suggesting that New Forestry, or New Perspectives, was little more than a shill for logging as usual. Nevertheless, by 1993, Forest Service chief Dale Robertson announced that the agency's scattered and limited New Perspectives effort was ready to come out of the cocoon, metamorphosed into ecosystem management. No longer was the new approach to be a scattered experiment: Robertson insisted that this New Forestry, now cum ecosystem management, would be applied through the National Forest system.

Environmentalists were immediately skeptical. "I applaud the Forest Service in using the rhetoric," said Jeff Olson, who directed something called the Bolle Center for Ecosystem Management at the Wilderness Society, a leading national environmental group that had long been in the thick of the spotted owl and old-growth forest controversy. "But I'm horribly disappointed in their failure to include any of the science or the logic of ecosystem management."

Olson told me then that he strongly suspected that the Forest Service was "taking what they're already doing and repackaging it under a new name—and we're left to wonder who they're trying to fool."

Indeed, in making the announcement, Robertson had pointed to seven "exceptions" to a major part of the scheme: a sharp restriction on clear-cutting. AFSEEE's Jeff DeBonis called the announcement "smoke and mirrors." He suggested that the exceptions, such as those that would allow clear-cutting in areas of insect and disease infestations, which are quite common in forests, or to "enhance wildlife habitat," could be so loosely interpreted that they could allow clear-cutting virtually anywhere.

"There are enough loopholes," he said, "to drive a logging truck through."

By the end of the year, the Clinton administration had replaced

Robertson with Jack Ward Thomas, the wildlife biologist who had led the series of owl studies. The appointment caused a stir in an agency where chiefs, in the past, had always come up through the ranks of trained foresters. Thomas was the first ever from the agency's scientific branch, and certainly the first "ologist." Thomas reasserted even more forcefully that, henceforth, ecosystem management would be the law of the national forests. Weeks after his appointment, I noted that many were skeptical that an agency, so long dedicated to the primacy of logging, could truly reinvent itself as an outfit dedicated to health and diverse ecosystems. Thomas bristled.

"I think that sells Forest Service people short," he said. "Until we've given our people a chance to view the world through different glasses, I think anyone's being premature to judge."

And maybe so. But I was growing skeptical myself. My own backyard, the forests of the Great Lakes region where I had done most of my growing up, had begun to recover from their own years of rape-and-run logging around the turn of the century. Trees were continuing to add wood fiber apace. And existing National Forest Plans indicated that virtually all of these woods—the Chippewa, the Superior, the Chequamegon, the Nicollet, the Ottawa and Hiawatha, the Huron, the Manistee, the Shawnee, and the Hoosier—would soon be subject to intensified cutting. But foresters in the region claimed that they were adopting New Forestry/Ecosystem Management fully. Several offered to show me their plans.

In southern Illinois's Shawnee, foresters indeed showed me a smallish, quite lovely native prairie and, across the road, a savanna dotted with oaks. Such ecosystems, interspersed with heavier forests, had once been common in the region. In northern Michigan and Minnesota, foresters showed me projects that created "openings" for wildlife, noting that many early successional species, ranging from grouse to some kinds of butterflies, often had as much trouble finding suitable habitat as late-successional (old-growth) species did. They showed me snags and logs left behind and maps showing how, in the future, they might aggregate their recently logged lands into larger blocks, thus keeping more mature growth (if not old growth) intact. It was on this same trip that Cleland and I visited the Kirtland's warbler habitat and the vigorous logging program there, and Jim Jordan's selectively cut northern hardwoods.

But something kept nagging at me. Even on a tour where the agency seemed clearly to be trying to put on its best face, virtually every ecosystem management project I was shown involved cutting down trees with chain saws. Why in the world, I wondered, didn't ecosystem management ever mean leaving the trees alone?

I put that question to Ken Holtje, who was ecosystem management team leader for the agency's Eastern Regional Office, in Milwaukee. It led to some hemming and hawing and an explanation that foresters could do a great deal through ecologically based logging.

"We can even create old growth!" he told me, referring to Jim Jordan's project in northern Michigan, the very project that Jordan had repeatedly insisted did *not* mean that old growth could be created.

And after a pause, Holtje added, "Now the objective is different. Now, producing wood products will be only one of our objectives. The chain saw is a just a tool."

None of which left me feeling less suspicious that at least some, and perhaps much, of the agency's "new" ecosystem management was not, despite some scientists' and Forest Service staffers' sincere efforts, going to be a new kind of forestry at all, but little more than the old wine of conventional farm-style silvaculture packaged with a new label.

Yet meanwhile, new discoveries were still to be made in the Andrews. And both within, and without, that laboratory forest, ideas were beginning to converge about what a really ecologically based forestry might in fact be—and what *not* truly changing forestry's conventional ways could mean in places far away from the shattered woods itself.

10

ON FEBRUARY 5, 1996, Gordon Grant began his day with a routine visit to the Andrews, or at least the virtual version of it, maneuvering his computer to the Andrews team's own World Wide Web site, where he quickly worked his way to a series of charts and graphs. A few years earlier, checking the array of monitoring gadgets for rainfall and snowpack moisture and stream flow in the Andrews would have meant a long trip into the mountains to read a series of paper graphs. But that has become a simpler matter these days. In fact, as I write this, I've just used the same Internet site to check the current state of the Andrews. (It is raining lightly on one mountaintop, but not at all anywhere else, and a few inches of snow remain on the highest peaks.) But easy or not, what Grant saw that morning stopped him cold. And after a long and astonished look at his screen, Grant hurried down the hall to find Fred Swanson.

Andrews team leader Swanson had distinguished himself early in his career with the group by describing a kinetic forest ecosystem—a place where mountain slopes and the living world upon them are in motion, slipping and slumping and sliding: a forest system moving fast, at least in the way that geologists, who normally think in terms of eons or epochs, think of time. It had become increasingly clear since Swanson's earliest studies that this moving forest did not move in a steady and uniform way. Although large changes might be detectable over centuries or decades, the most dramatic changes themselves appeared to occur suddenly, in bursts. The most significant moves seemed to come in pulses—Grant calls it a kind of punctuated equilibrium (after a term developed by evolutionary biologists to define nature's patterns of sudden pulses of speciation and extinction, followed by long periods of more subdued evolutionary activity).

Another related issue had become clear: although the biggest of these landscape-moving pulses were "frequent" in geologic terms, they were so widely spaced in human terms that a geomorphologist might go through an entire career without experiencing one. And certainly, if the forest ever did begin to move dramatically, it would be as big an event for a forest geomorphologist as, say, a Mount St. Helens eruption would be to a volcanologist. And that's precisely why geomorphologist Gordon Grant was in a rush to find geomorphologist Swanson.

Two days earlier, on February 3, a mass of unusually warm air had begun to accumulate in the western reaches of the Pacific Ocean. The immense warm-air mass then consolidated and moved eastward, inhaling seawater. On February 5 the sodden air mass reached the Oregon coast. It slid up the Coast Range, which fronts the Pacific, and then moved across the Willamette Valley to the east, then up the higher Cascades, cooling as it rose, and, inevitably, releasing its water, letting loose almost as if a gargantuan bucket had scooped up part of the ocean and was dumping it almost at once. A deluge of biblical proportions began. In subsequent days, Portland television stations issued astonished and breathless reports about the devastation wrought by simple rain: rivers raging out of the Cascades twelve feet over flood stage, houses in Portland flooded to their windowsills, rural houses literally swept away as muddy landslides roared down mountain slopes, entire freight trains flipped off their tracks by rushing debris, and, over all, millions of dollars of damage in the disaster area the state had become.

The glowing charts on Grant's computer screen that morning informed him that a deep snowpack still remained in the forest's highest slopes. A pressure gauge nestled inside a plastic pillow stuffed with ethylene glycol antifreeze showed that snow on the mountaintops held the equivalent of twenty-three inches of rainwater, about ten feet of average-density snow. Near the bottom of the slopes lay another two feet of snow—or another six inches of water. Altogether, that amounted to an enormous amount of wet snow just on the verge of melting.

That much snow normally would melt away gradually, over weeks. But what Grant saw on his screen that morning, combined with what he knew about climate dynamics in the mountains, hinted that a rapid, wet drama was under way: enormous quantities of warm rain pelting onto the already saturated spring snowpack was forcing an almost instant melt. Just outside the Andrews, streamflow gauges showed that the McKenzie River was rising at a rate vastly beyond anything Grant had ever seen before: the river's flow was increasing by an almost incomprehensible, torrential one thousand cubic feet per second every hour.

In combination, those facts told Grant one big thing: the Andrews ecosystem was about to move, and in a big way.

Grant recalls first running into Swanson as an undergraduate in the early 1970s at the University of Oregon in Eugene, near the start of the Andrews project, when Swanson taught a class called Forest Geomorphology.

"It was the first time I'd ever heard the word *geomorphology*," Grant says. Although he says he was intrigued by the topic that would eventually become his life's work, Grant did not follow a traditional path from undergraduate to graduate school to science ("I basically majored in rivers at Oregon," he says) but rather spent a dozen years after college as a western river guide. "But I found myself sitting in boats, asking, why does this river look the way it does?"

By 1986, an intensifying curiosity about how rivers form and change had led Grant to a doctoral program in fluvial geomorphology at Johns Hopkins University in Baltimore, and by the early

1990s he found himself heading back west to Oregon for a research job with the Forest Service, as a member of the Andrews team. At the core of his fascination with rivers, he told me, is what Oregon writer John Daniels has called their "ghostly mathematics."

"A river flows not just downstream in two dimensions, but in three," he told me one day in his office, doing a sort of tilting dance in his office chair as if to demonstrate the flow. Rivers meander, for instance, with those in the flattest country the most serpentine, and with a remarkably rigid mathematical correlation between the width of the channel and the length of each meander: each meander wavelength is ten to fourteen times the channel width, and the same rule applies to other flowing masses—from the streaming currents of the ocean to the jet stream high in the atmosphere.

"It's easy enough to observe something like this," he says, "but to explain it takes a blackboard full of differential equations. When something like that storm occurs, people want me to be able to tell them what it will mean for downtown Portland, and they get annoyed if I tell them there are too many variables, that I don't know enough yet to tell them, that I'm an expert in a science that maybe isn't in its infancy, but maybe only its early adolescence."

On that rainy February day, Swanson and Grant both faced a singular opportunity to confirm ideas that could advance that knowledge, one they might never again have in their careers. The two dropped everything, rounded up three equally curious colleagues, requisitioned a Forest Service truck with four-wheel drive, and headed south on Interstate 5, then east on State Highway 126, meandering along the McKenzie upstream toward the Andrews. Rain was pelting the truck, pouring in sheets off the windshield, the headlights barely penetrating the deluge in what should have been full daylight. It was a lousy time to be heading up into the Cascades. But the small group of five finally reached the Andrews in the dim light of late afternoon. As they drove through the roads on the site, not much seemed to be happening out of the ordinary—a small slide of debris here, water pouring down a hillside and across a section of road there. Still, the streams were swollen and raging, and when the scientists climbed out of their truck near lower Lookout Creek, they could hear big boulders rolling and banging on the stream bottom. "Not knock knock knock," says Swanson, "but kawhoomp kawhoomp kawhoomp."

By the time they had completed a preliminary check of the experimental forest, darkness was approaching. The group left for an early dinner at a nearby restaurant. It was dark when they returned. "When we got back to the Andrews," says Swanson, "all hell had broken loose."

No longer just water, but what Grant called a brown ooze—a slurry of water and soil—was pouring over the road in the beam of their headlights. A boulder "the size of a Volkswagen" blocked another road. "We knew," Grant says, "that the hill slopes had begun to move."

A kind of stream was actually flowing down the road, partly contained by berms of plowed snow on either side, and filled with mud, chunks of wood, and other debris. Swanson and Grant quickly surmised that there had been a landslide further up the valley and carefully they drove a few hundred yards further up the road until the truck's way was blocked by debris and deep gullies that had formed in the roadbed. The group clambered out of the truck and, linking hands (in part because Grant is night-blind), worked their way further up the road on foot, by the beam of their only flashlight. Perhaps a quarter mile from the truck, they reached the mouth of a tiny creek simply called Watershed Ten, where, indeed, water and landslide-debris were pouring onto the road.

On a landscape where giant boulders were rolling down mountainsides, there was little to recommend working in the dark, so the group retreated to spend the night at the Andrews headquarters complex. At first light, the scientists, outfitted in bright yellow rain suits, were back on the scene, standing on a bridge over a Lookout Creek that had turned into a torrential river. Rain still poured from the skies. Given the volume it was carrying, Lookout Creek would have looked like a major, roaring whitewater river if it hadn't been for the fact that it was carrying not only water, but a thick slurry laden with mud and sediment. The sturdy wood-beam Lookout Creek bridge was shuddering beneath their feet. As they stood there, most of the bole of an immense old-growth tree appeared just upstream.

"Oh my God!" shouted Grant. "It looks like . . . it *is*, it's an aircraft carrier. Look at that thing!"

The huge log raced toward them, but then shot under the bridge; it took a sharp left turn with the foaming current and was gone. More logs and parts of logs pelted down the river, along with giant, bobbing root-wads, tangled masses of thousands of pounds of root torn from the soil. Swanson and Grant had been theorizing about the movement of huge pieces of woody debris on the landscape and in streams for years but now, for the first time, they were actually seeing the pieces in motion.

The rains finally subsided, but in the next few weeks Swanson and Grant and a score of colleagues who descended on the Andrews found an ecosystem transfigured by the storm. The streams had seen dramatic changes: gravel bars and alluvial fans where none had existed, old gravel bars and fans gone. Massive logjams that had remained in place for decades had broken up and moved; new jams had appeared. The storm also had profoundly altered the living communities on the stream banks. Big logs like Grant's "aircraft carrier" roaring down streams had swept away whole stands of alders and other streamside trees like giant scythes. And scattered all along the creek were the end points of massive landslides—"debris torrents"—of rock and mud and even whole old-growth trees that had pummeled their way down mountain slopes at speeds of up to twenty miles an hour, delivering tons of sediment and organic matter to the waterways.

Later, Grant would characterize the process of geomorphic change in the ecosystem as "decades of boredom, punctuated by hours of chaos." It is during those few hours of chaos that much of the shaping, re-forming, and cyclical disturbance of the forest ecosystem, and particularly its streams and riparian (streamside) zones, takes place. "The shape of the channel, the position of big logs, it's all reshaped in a twelve- to twenty-four-hour window when the channel is in full flood," Grant says.

All of which may be profoundly exciting to a curious geomorphologist (indeed, on videotapes of the storm, the usually reserved Swanson seemed to be nearly jumping out of his rain suit. "Whooaaaa, RIP CITY!" he shouted as one big root mass roared down the stream). But although the scientists are still in the process of trying to sift through all their data about the storm's effects, important patterns already have emerged.

The storm caused a renewed debate in the region about the role played by clear-cut logging and road building in the upstream reaches of Pacific Northwest rivers in the devastating flooding of downstream cities during big storms.

Not since 1964 had a massive storm visited the region. And in fact, the 1964 storm had been less intense in the Andrews. Curiously, says Swanson, that smaller 1964 storm had led to *more* streamside disturbance than the more intense 1996 one. The only apparent explanation for that, he insists, is that something had changed in the Andrews ecosystem itself in the interim, something that created a sort of protective buffer that reduced the latter storm's impact. That protective buffer: a forest freer of the effects of road building and clear-cutting.

Despite the fact that it has remained largely old growth, the Andrews *had* been a working forest before the 1964 floods. For fifteen years before that flood, the Forest Service had been building a network of roads through the experimental forest, just as it built roads in its working forests elsewhere. And all during those fifteen years, loggers had been making experimental clear-cuts in scattered parts of these woods.

In those early years, the Andrews did not exist as a place to study how old growth functioned ecologically, but as a place to try to figure out how most efficiently to cut it down and regenerate an "improved" version. But since the beginning of the Andrews whole-ecosystem study, there had been a quarter century of almost no further human disturbance in the ecosystem. The scattered old clear-cuts had begun to heal; old roads had been abandoned; new road building had ceased.

Does that prove that a more intact forest is a better buffer against giant storms? Not definitively, says Grant. Nor, he suspects, will he or Swanson or any other scientist in the near future ever be able to prove such a thing beyond any doubt. Such big storms are so infrequent that it would take dozens of lifetimes to collect and compare enough data for unshakable statistical proof. Even attempting to analyze data from a handful of more frequent but less intense storms that occur, say, once every ten years on average, would be tantamount to trying to predict the outcome of a presidential election by polling only a handful of voters.

The timber industry and its supporters in the region have, in fact, long insisted that logging does not significantly increase flooding during big storms. But even before the 1996 storm, Grant and Andrews scientist Julia Jones had set out to examine the effects of smaller and more frequent rainstorms—the kinds that would provide enough data points to draw more incontrovertible conclusions. Looking at a sampling of small Oregon watersheds, they found dramatic differences between those that had been logged recently and those that had not. They discovered that within five years after a cut, peak flows in nearby streams typically increased an average of 50 percent during even the smaller storms. And, they say, their analysis shows that as long as twenty-five years after a clear-cut, flows remain 25 to 40 percent higher. Even more startling, they found that logging as little as 5 percent of some watersheds could increase peak flows by anywhere from 10 to a stunning 55 percent.

Jones says, "We think the mechanism is the road network transmitting the water, not simply the clear-cut area. The road network can intercept water that would be moving slowly in the soils and put it into roadside ditches or road surfaces where it goes much faster, and that's largely what accounts for the increase in the peak flows."

Jones also notes that storms big enough to turn on this road effect tend to occur at a particular time of year—winter, when the soils are already sodden and vegetation is relatively dormant (and thus not taking up water through root systems). Whether the Jones and Grant study can be extrapolated to reflect the effects of monster storms like the 1996 deluge remains an open question, since their study was necessarily limited to the kinds of more moderate, if still big, storms that might occur from one to five years. But since the rare, immense, 1996-style storms also usually occur during winter, Jones and Grant have openly speculated that the extreme storms would behave similarly.

"We see no reason for bigger storms to behave differently," the two scientists wrote in their report, published in 1996 in *Water Resources Research*.

Some implications are obvious. The 1996 Oregon storm caused millions of dollars of damage to the built environment alone—in the form of destroyed or damaged roads, railroads, houses, and commercial buildings. Another is far less obvious, but in terms of diminishing natural biodiversity, and natural-resource economics, deeply troubling: excessive flooding, and the increase in sediment in streams that accompanies it, can devastate spawning habitat for some of the region's most highly prized and economically important species of fish.

If any organisms in the Pacific Northwest are more famous than the region's giant trees (and, more recently perhaps, the spotted owl), they are the handful of species of anadromous fish, all members of the trout-salmon family, that abound—or once abounded—in the region's adjoining seas and, for part of each year, in its rivers. (Anadromous fish are those that live most of their adult lives in the saline waters of the ocean, but return to fresh water to spawn.) By the time of the great 1996 storm, stream and fisheries scientists working in the Andrews and elsewhere in the region had already begun to warn that extensive logging and, especially, road building in such a mountainous region could slowly but steadily devastate salmonid reproduction and survival.

As recently as the 1960s, the annual migrations of these often huge salmonids—cohos, chinooks, sockeyes, pink, and chum, and steelhead trout—were among American nature's most spectacular events. In their breeding seasons, giant and silvery salmon and trout would work their way by the millions far up the Pacific Northwest's rivers, in order to spawn.

The salmonid migrations themselves are a marvel. Much of the actual mechanics remains a deep mystery of nature. Chinook, or "king," salmon, for instance, which reach an average size of about twenty-five pounds as adults, remain out at sea for from three to seven years. They may travel as far across the Pacific as twenty-five hundred miles from the stream in which they were born—nearly the width of the North American continent. And yet, somehow, they find their way back to reproduce in the very stream in which they were hatched. What they need from the stream is nothing complex: clean water, rich in oxygen, an open passage through that water, and the silt-free beds of fine gravel on which they have spawned for millennia.

But in the latter half of the twentieth century, these magnificent fish have found less and less of that. Annual runs of coho salmon in Oregon and northern California, for instance, have declined, at this writing, from 1.4 million fish to fewer than thirty-nine thousand, a decline of more than 97 percent. Coho salmon, in fact, are already completely gone from more than half of their historic ecological range.

Damage has come from four key areas, the "four H's" in the parlance of northwest fisheries science: outright habitat destruction, hydroelectric dams on their migratory rivers, intense harvest of rare stocks, and competition from hatchery-raised fish. A 1991 report by the American Fisheries Society reported that 214 of about 400 "stocks," meaning genetically unique races of salmon and trout in the region, were moving fast toward extinction. Fully 106 stocks, said the report, were already extinct.

In economic terms alone, the precipitous and rapid collapse of the region's salmon fishery has been horrific. Over three decades, the Pacific Northwest had lost some seventy-two thousand jobs in the salmon-fishing industry, and by 1998 it had lost an estimated 1.5 billion dollars per year in annual wages and profits, according to Glen Spain, the Northwest regional director for the Pacific Coast Federation of Fishermen's Association.

Fisheries ecologists on the Andrews team had long been convinced that protecting remaining fish stocks, and any hope of rehabilitation for those in steep decline, meant protecting the quality of streams.

Salmon spawn in beds of gravel, first fanning away fine rocks to build bowl-like nests called redds, laying and fertilizing their eggs, and then fanning a layer of lighter gravel over the eggs as protection from predators. Logs in streams in old-growth stands slow the flow of water, which in turn help create suitable gravel beds by allowing suspended fine rocks to drop onto the stream bottom in the quiet eddies behind the logs. The salmon's life cycle is finely evolved to interact with the physical structure that the complex of logs and gravel provides: in good habitat (which once abounded in the Northwest) the fish find gravel beds with stones small enough that they can move the tiny rock particles with a fanning action from their bodies, and at the same time large enough to fit only loosely together, which allows water to flow around the eggs, exposing developing embryos to a rich supply of oxygen.

If silt and sediment clog the gaps between bits of tiny rock, the embryos become starved for oxygen and perish. The Andrews "stream team's" research long before the 1996 deluge showed that even in more normal rainstorms, the cumulative effects of logging and, especially, road building could increase the flow of sediments to salmonid streams by anywhere from ten times to one thousand times.

The situation is not simple, however. Jim Sedell and a group of colleagues discovered in the early 1980s that fish stocks in some colder streams actually can *increase* in the first decade or two after logging of old growth, because any problems the logging had caused in the streams is offset temporarily by the fact that, with the canopy gone, sun pours into a stream, promoting the growth of algae, which in turn causes populations of aquatic insects to boom, which in turn provides a rich source of food for young salmon. However, after perhaps two decades of such abundance, and long before the forest is mature enough to log again, the stream will lie once again in shade, and the salmon populations will quickly crash to levels far below those supported by streams running through old growth. That effect, in any case, applies to the coldest waters, generally small headwaters creeks high in the mountains. In somewhat warmer reaches—generally larger streams at lower altitudes, the very streams with habitat to support the most diverse array of fish—the removal of the shading old-growth canopy can mean that waters overheat. In one study of streams in the Oregon Coast Range, a stretch of recently logged stream soared to 84°F in midday sun, an extremely lethal temperature for salmon; under a maximum shading canopy, it would have reached only 54°F. Particularly in the southernmost, and thus warmest, part of the salmonids' range, such as southwest Oregon and northern California, streams in cleared areas blasted with sunlight routinely rise beyond the mid-70 degrees in midsummer, deadly temperatures for the fish.

As striking as those discoveries might be, they pale in comparison to what the scientists found during the 1996 flood. As they had predicted, ditches and roads served as superhighways for flooding waters during the deluge. When excessive runoff accumulated in roadside ditches, it blew out culverts or suddenly roared over mountain roads, washing out entire sections of road and then carving away the mountain slopes below, carrying big chunks of the landscape with it. Gravity drove the process, with rushing torrents of water and rock and muck flowing to depressions on mountain slopes, then rushing en masse further down. Because the soil that had supported some trees was water-blasted away, those trees crashed to earth and joined the roaring debris torrent, a torrent accumulating more power and speed until it finally crashed through streamside vegetation and into a stream as part flood and, thanks to the huge trunks and boulders that had joined the torrent, part giant plow, scouring the stream bottom and banks and ripping out riparian trees and shrubs. Grant's calculations suggest that more sediment entered streams in the few hours of the single 1996 flood than had entered in the previous three decades!

But not all streams suffered similar sedimentation. On slopes where there were few, or no, roads, there were no roadways and ditches to funnel water, no roadbeds to collapse. And along streams lined with wide expanses of giant old-growth trees, those trees often acted as barricades, dikes against the giant debris flows.

The real problem is not flooding and debris torrents or even sediment. Swanson notes that huge storms have always visited the region, and that long before logging arrived, those storms served as a natural disturbance, much the way fire serves as a wholly natural disturbance in many types of forests. Even on landscapes untouched by development, some landslides and debris torrents still inevitably occur. If limited to a scale that nature can accommodate, debris torrents might well help naturally "restart" the stream ecosystems by, say, providing new logs to the stream as fish habitat, or by scouring old deposits of sediment from gravel beds.

As ecologist Elliott Norse has written, "Recent findings suggest the model that managers should emulate to improve salmon runs. It is the streams produced by ancient forests."

Decades of boredom, punctuated by hours of chaos, Grant had called it.

"Yeah," Swanson says with a small smile when I mention that. "And the point is, a lot of our most important work is necessarily focused around those decades of boredom. I think part of the reason I was so *pumped* during the flood wasn't so much that we were formulating new hypotheses from the stream bank, but that we were watching all these other hypotheses from twenty-five years of work being confirmed."

I had told Swanson during an early visit to the Andrews that one of the issues that most intrigued me was the sheer process of discovery as a team of scientists sets out to understand an ecosystem like the old-growth forest for the first time.

He has been thinking a great deal about discovery. The process of trying to understand an ecosystem, he says, is not one typified by "eureka moments at streamside," but rather of year after year of slogging along, collecting mountains of data, trying to sift through them and make sense of them.

"To the extent that we've made important discoveries here," he says, "it all goes back to the idea of a seedbed—the commitment by a core group to doing long-term science, and the commitment to a place."

Art McKee, botanist and Andrews site manager, was leaning against Swanson's office door, nodding, as Swanson said that. Commitment to the Andrews as a place in the early days, McKee recalled, meant begging for funds (after the initial International Biological Programme grant monies ran out) and days, weeks, sometimes months of Spartan living in the "ghetto in the meadow," the ramshackle collection of shacks and trailers that were the Andrews headquarters.

"I remember a scout-type wilderness camp I went to as a kid that was a lot more luxurious than the old Andrews," McKee says. Today, the Andrews facility is the third largest forest research center in the United States (just after the Corvallis lab on the Oregon State campus where McKee and Swanson's real-world offices are located; the largest is the legendary Forest Products Laboratory in Madison, Wisconsin, started by Aldo Leopold and now working on such esoterica as recycling degraded waste paper into plastic polymers) and one of the handful of largest forest laboratories in the world. Both Swanson and McKee say they worry about that size and status sometimes. For all of the relative comforts offered by the trim dormitories and office and research buildings at the site, they are concerned, they say, that the almost Bohemian passion of the early years has been drained from the endeavor.

Yet the site remains a remarkable "seedbed." The long, slogging, often tedious days of basic data collection and pulses of new discovery from its early days now seem to be providing a launching pad for entirely new kinds of investigations.

One example: Bob Griffiths, a microbiologist who is technically an emeritus professor at Oregon State, but who retains an office on the campus and continues to advise a handful of Ph.D. students, has become fascinated with the Andrews group's work on subterranean fungi. Griffiths, a cheerful, broad-faced man with a hearty belly laugh, spent virtually all of his career studying microbes in the oceans. These days, he is given to roaring to his own research sites in the Andrews on a Harley Davidson, and has begun to uncover wholly new secrets of the world of forest mushrooms and truffles and mycorrhizae.

Griffiths, his graduate students, and a handful of colleagues have found, for instance, that great mats of fungi spread over not only the Andrews and similar old-growth forests in the Northwest, but most mature coniferous forests they have sampled, from the border of the Arctic tundra and into California. Those mats can cover as much as one fourth of the forest floor and, where they are dense, can constitute as much as half of the total biomass of the forest soil. But far more important, Griffiths's analyses of changes in the chemistry of the fungi and the soil have convinced him that these great matted networks of fungal hyphae appear to perform roles in the forest ecosystem beyond their known role as extensions of the root system.

"A decade ago, it would have been heresy even to talk in these terms," he says. "In the very recent past, everyone agreed that the function of mycorrhizae was pretty much limited to taking up nutrients and water—that the plants supported the mycorrhizae and in turn got a very extensive network of fine roots."

But now, he says, new discoveries suggest that the fungi also exude enzymes that are critical for part of the decomposition of voluminous organic debris in the soil that are among the most difficult to break down. This so-called humic material is actually the leavings of earlier stages of a decomposition process that first breaks down the most immediately usable substances, like sugars in a leaf (a process that can occur in only hours after the leaf falls), to a stage where such organisms as bacteria "clip" away simple starches (a process that can occur over, perhaps, a year), to a third stage where soil decomposers slowly break down more tenacious molecules like those that form the tough, gluey lignin that binds cells together to make wood (which might occur over a decade or more). In the end, those stages of decay leave an abundance of humic molecules—large, tenacious polymers that are difficult to break down. Conventional scientific wisdom had long held that this wealth of humic matter breaks down only if it is processed first by specialized bacteria, and then through a slow and complex biochemical food chain. But Griffiths's work suggests some of the mycorrhizae that mat together in older forests can simply extract elements like nitrogen and phosphorus from the humic matter directly: an analogue of getting pure food intravenously rather than through the complex and messy process of chewing and digestion.

"In effect," he says, "they short-circuit the whole food web."

Meanwhile, Griffiths and his coworkers have found that some of the mycorrhizal species are virtually rock splitters—that these fungi can dissolve stones in rocky soils and pump their embedded minerals into the forest ecosystem. Andrews scientists Dave Perry and Michael Amaranthus have shown that mycorrhizal density plummets after forests are clear-cut, and now Griffiths suggests that their missing role in mineralizing rock and freeing nutrients from humic matter might well explain why some replanted forests are simply unable to thrive on some sites.

"The bottom line for forest restoration is, how did these mats get there in the first place? If they're gone, how can we get them back? If you cut down the trees, can they maintain themselves somehow for a while on the roots of some alternate plant host, or can they function independently for a while? We don't know yet, but it gets to a big question: what the hell should you really be doing after a site is logged or burned? If you're whacking the site with herbicides to keep down weeds or brush, are you also whacking the alternate hosts the mycorrhizae need to survive?"

Team member Tim Schowalter, an insect ecologist, has been collecting and trying to pin down the role of the abundant insects, spiders, and other arthropods that live at various levels in the forest canopy. Most species are predators. That fact, he says, strongly implies that these tiny animals are a major, and likely *the* major defense old-growth trees have against defoliating herbivorous insects. In his recent canopy studies, too, he has begun to find staggering numbers of oribatid mites—not just a host of new species of mites but even a few new and distinct genera (coyotes and wolves are different species of the genus canis; wolves and tigers are different genera). Many appear to be feeding on the nutrient-rich "scuz" of bacteria and other tiny organisms on needles and cycling those nutrients back to the forest as they defecate or die. And he is also fascinated, he says, by the ecological role played by the defoliators that *do* manage to survive in the forest.

"We think of these insects as pests," he says, "but we're learning that they can play extremely critical roles in the *health* of the forest. Even when they defoliate part of a tree and kill a branch, or when they kill whole trees, they often do it for a good ecological reason. We're learning that they're better managers of the forest than we are."

As an example, he says, it now appears that defoliators that successfully devour and destroy needles on the lowest, most shaded branches of a growing, maturing Douglas fir tree actually serve to prune branch and needle systems that are draining more resources from the tree than their weak photosynthetic capacity is returning. Shaded and weakened, they become susceptible to pests. But once removed by the action of those defoliators, the pruned branches not only no longer parasitize resources from the tree, but also no longer present an accessible offering of ready fuel to a fire moving along the ground. The fact that defoliators strip away the lower branches that could serve as fuel ladders for fire to move into the high canopy, he says, is one of the reasons old-growth Douglas firs are so fire-tolerant.

Similarly, he says, an apparent scourge of defoliating spruce budworm that raged through the driest, low-lying forests on the east side of the Cascade Range (that is, on the other side of the Cascades crest from the Andrews) between 1987 and 1992 could more sensibly be seen not as a disaster but as "an ecosystem trying to tell us something." Fire had been suppressed in this naturally fire-prone ecosystem for years. By the late 1980s, trees that had been replanted after logging, many of them Douglas firs poorly suited to these drier soils, were competing fiercely for limited water. The budworms were merely killing the least hardy trees.

"The budworms were seen as the problem—killing so many trees. But if you really look at it, the budworms weren't the problem, they were the *solution* to the problem."

The most important message Schowalter has to offer is that sorting out the role of arthropods of the old growth, and their true complex of roles in the ecosystem, remains one of the most enigmatic problems scientists still face.

"We literally do not know what to expect next. We do know that every time we look, whether it's in the soil, or in the canopy, or in rotting logs, we find fantastic levels of diversity. When we looked more closely at log decomposition, we found double the number of wood-associated arthropods than we even knew existed before. In the Andrews, we've got a unique forest that everyone can appreciate because of the size of those trees. But we haven't even begun to appreciate the diversity of life-forms just in that one place. We're nowhere even close to understanding what role they all play, but we have every reason to believe they play very, very important roles we don't presently have a clue about in terms of the health of the forest. We do know we find different complexes of species in old forests than in young ones. If we lose some of this diversity in our manipulations of these ecosystems, we could be losing elements that are critically important under different environmental conditions. I can't think of a more powerful argument for maintaining biodiversity in these forests."

Schowalter's research techniques are decidedly on the low-tech side: he often sends assistants out to collect whole branches, which they shove into plastic garbage bags. He refrigerates the entire branch to put the canopy arthropods in a torpid state, collects them with a tweezers or a small paint brush, and studies them under a microscope.

Dramatically, on the other end of the technology spectrum, remote sensing specialist Warren Cohen has been using the team's detailed seedbed knowledge of the Andrews to study first the experimental forest, and then much of the Pacific Northwest—and from hundreds of miles above the earth. Cohen's tools are mounted on orbiting satellites, like the National Aeronautics and Space Administration's LANDSAT satellite, and a French satellite system nicknamed SPOT (Système Probatoire d'Observation de la Terre).

Although most of us are familiar with the kinds of weather satellite photographs that appear on the evening news, the satellite pictures of most interest to Cohen typically employ specialized, extraordinarily sensitive sensors that detect faint flickers of electromagnetic waves invisible to the human eye. Cohen has found the sensors discriminating enough that he can, say, distinguish on a satellite image between needle-leaved and broad-leaved trees, and even between old growth and young growth, between open clearings and brush.

As he explains it, even a photographic film that records images only in the spectrum of light visible to humans can distinguish such factors as whether a stand of trees, seen from above, is coniferous or broad-leaved, or old growth or young. The flat surfaces of broad-leaved trees reflect more visible light, and seem lighter, greener. Dense and complex masses of needles on a conifer absorb more visible light and thus appear dark, sometimes nearly black. Old-growth stands are typically filled not only with their characteristic tall trees, but many gaps, dappling them with large, light-absorbing shadows. The satellite sensors offer an abstract but fairly precise view, receiving an enormous array of electromagnetic signals at wavelengths ranging from infrared to ultraviolet, translating those signals into digital form—numbers—and beaming them back to Earth to be processed by computers into usable data. When preparing maps and images based on that data researchers choose which color to assign to which electromagnetic signal (dark green, perhaps, to a signal associated with old growth, tan for a signal associated with a clear-cut, blue for a signal associated with water. The color choices, however, are arbitrary; the color for water could as well be pink.)

It was the extensive and detailed maps of existing vegetation in the Andrews that allowed Cohen to proceed in such detail at all. He built new and complex statistical models based on the raw data he had received from the satellites, combining his research team's knowledge of how forests reflect light with knowledge of existing forest characteristics in the Andrews.

"We already had the best field inventory data you could hope to put together," Cohen says. "It really was just a matter of comparing that spectral data with the existing field data. From there, we were able to go on with an initial project to make a kind of simplified map of Oregon's forests from Eugene to Portland, from the Willamette River to the crest of the Cascades." Now his maps include all the lands in the state west of the Cascades crest, a map bejeweled with colors representing an extraordinary array of details, ranging from the relative ages of stands of trees to whether or not entire stands of replanted forest are growing vigorously.

A key advantage to being able to analyze ecosystems from space: a researcher like Cohen can analyze vastly more territory in a matter of days than he otherwise could in several lifetimes. As this book went to press, that ability to look wide and deep had Cohen in the late stages of confirming a new and disturbing discovery: in vast areas of the Northwest, significant portions of industrial-style tree plantations seem not to be growing the trees that were planted there after old growth was logged out, but only brush. These are the very plantations that were to have replaced the region's supply of timber once most of the old growth was cut down by early in the twenty-first century. And they are, to a large extent, the supposed future of the American lumber and wood fiber supply. The findings still need to be confirmed, but if planned, detailed "ground-truthing" checks confirm the results, it is a bombshell discovery.

Trees seem to be regenerating poorly in two specific areas, Cohen says. "The hottest and driest south-facing slopes are generally not doing well, and at higher elevations, the north-facing slopes are not regenerating either." The precise reasons? No one knows yet. But since much of the already clear-cut lands are industrial holdings—functional tree farms—the future timber supply could plummet billions of board feet below projections. Fully 20 percent of the replanted plantation lands, he says, appear to be coming up only in brush.

"It really does cast some doubt," declares Cohen, "on the mentality that says, 'Let's get rid of all that unproductive natural forest and replace it with plantations that'll grow like crazy.' "

Earlier and more limited versions of the maps gave Cohen and a small group from the Andrews team a unique opportunity to settle a critical debate about the roles played by old growth, and logging, in climate change. Some timber industry and forestry school proponents of logging out old growth had suggested that clear-cutting followed by replanting would be good for the planet, in that it would replace stands of old trees, which were adding little or no new growth, with vigorously growing young trees. The young trees, in turn, would inhale vast amounts of carbon dioxide, easing the effects of global warming.

But in a 1996 paper published in the journal *Bioscience*, Cohen, Mark Harmon, David Wallin, and Maria Fiorella combined detailed maps with an analysis of carbon uptake and release by living and dead trees and handily refuted that notion. They noted that the logging proponents were correct on one major point: living old growth did consume less carbon dioxide than younger, more vigorously growing stands of trees. But the trend reversed when the scientists added to their formula the carbon released by everything from rotting woody root systems on recently logged sites, to branches and twigs either burned on the site or left to rot on the ground, as well as the carbon released by the manufacture of lumber or paper, as well as the additional carbon released as these wood products were eventually burned, or as they slowly decayed.

11

WHAT SURPRISES AND WHAT NEW DISCOVERIES the Andrews seedbed will yield in coming years, or decades, or even centuries (if Mark Harmon's optimism about his two-hundred-year log study is justified) remains anyone's guess. But it seems undeniable that a quarter century of discovery here is already challenging, and changing, fundamental assumptions of the science of forestry. It remains to be seen how long it might take some entrenched U.S. Forest Service managers to fully embrace more ecosystem-based approaches. But its New Perspectives debacles notwithstanding, signs abound that forestry is already moving down a more ecologically based path. Some instances of dramatic change already are under way in the agency: notions such as imitating natural disturbance patterns and managing large landscapes for diversity are on many lips. (In contrast, when the three Wisconsin scientists who appear in this book's preface urged the Forest Service in the late 1980s to protect large islands for biodiversity, the idea was rejected on the basis that it was "only a theory.") And at least as important, new ideas and approaches are finding their way into plans to manage forests more soundly beyond federal lands.

One of the most dramatic developments has come in the Northwest itself. In the 1992 election, candidate Bill Clinton carried the Pacific Northwest with a promise that through a process of consensus and compromise, he was going to find an answer to the long, grinding, endangered-species-versus-jobs (or loggers-versus-environmentalists) controversy that had raged for years. Already, most timber sales in the spotted owl range on federal lands were frozen by a court injunction related to the owls' threatened classification under the Endangered Species Act. Mills that specialized in sawing up old-growth timber were starved for supplies, and neither side in the debate seemed ready to give in. True to his word, Clinton in the first weeks of his presidency staged a showy summit between key players in the dispute and his own highest-ranking aides and advisors, including Vice President Al Gore, and then appointed a high-level scientific committee to come up with a range of possible answers. Although once again headed up by Forest Service wildlife researcher (and soon to be chief) Jack Ward Thomas, the scientific committee was not only heavy with scientists from the Andrews team, but suffused with concepts and ideas it had helped developed. In terms of the protection of nature, the plan was to go far beyond any previous efforts. It was not a spotted owl protection plan, but a spotted owl, and marbled murrelet, and (especially) salmon and, generally, a biodiversity protection effort that aimed to protect ecosystems for well over a hundred species thought to be now, or soon, imperiled, including, perhaps remarkably, a panoply of old-growth-associated insects and amphibians and even fungi. The team came up with eight different options for forest ecosystem protection and logging: even the most lumber friendly of those plans called for enormous reductions in federal timber sales in the region: an 80-plus percent cut from 1980s levels of four billion board feet to about seven hundred million. Convinced that such a plan had no hope of surviving politically, Thomas recruited Jerry Franklin to rapidly work out yet another approach. Dubbed Option 9, and ultimately adopted by the Clinton administration, the plan was a true compromise: it pleased no one. But it was (just barely) enough to convince Federal District Court Judge William Dwyer to lift his long injunction on logging in spotted owl country.

At the plan's heart was this: remaining old growth on federal forest lands in the Northwest would be divided into categories—about three fourths (partially) would be protected lands in old-growth reserves, designed particularly around maps showing overlaps between potential spotted owl nesting habitat and key salmon headwaters streams. Logging of younger trees, under eighty years old, would be allowed in these reserves provided it actually enhanced old-growth characteristics (for instance, by creating gaps). And "salvage" logging of dead or diseased trees would also be allowed sometimes. Elsewhere, critical buffers would be maintained around streams. Another fourth of the present old growth would be placed in giant (eighty-thousand- to one-hundred-thousand-acre) adaptive management areas, where foresters could experiment with New Forestry or other cutting-edge ecosystem management approaches.

Although some environmentalists have decried any further cutting in the relatively few bits of old growth that remain, few have objected, or would object to the dramatic transformation the Clinton forest plan has brought to one of Oregon's most productive forests (of both timber and biodiversity)—the Siuslaw. It was in the Siuslaw, which runs along the Coast Range, that Randy Molina and his colleagues had brought me to hunt for truffles. And it was the Siuslaw that as recently as the 1980s was the apple of the timber cutter's eye. In the timber boom years of the 1980s, some 350 million board feet were being logged out of the Siuslaw annually, generating some 1.8 billion dollars in revenues. Under the new Northwest Forest Plan, almost nothing—five million board feet—is being cut each year. But forest supervisor Jim Furnish says the value of the forest to biodiversity, notably salmon reproduction, is at least as high as those former billions in log sales, and based in large part on the Andrews forest and stream studies, he is aggressively returning his forest to wildness: of the forest's former spaghetti works of twenty-four hundred miles of roads he has, at this writing, closed or "retired" about twelve hundred miles and is working on shutting down another seven hundred—or about 80 percent of the total. Instead of supervising timber sales, his staffers are supervising crews that are removing culverts, returning slopes to normal grades, and cutting "water bars"—

small drainage channels, every few feet along any remaining roadbeds, to promote the kinds of drainage patterns that a natural mountain forest would exhibit, especially the tendency of water to flow
down the slope in a sheet, rather than as a gushing temporary stream.

In Michigan, the Forest Service settled a lawsuit in the early 1990s
with the Sierra Club and Wilderness Society over the lack of wilderness and old growth (to protect biodiversity) in its two forests in the
Lower Peninsula. The resulting settlement: ecologist Dave Cleland
designed a flowing reserve of core areas and river-based travel corridors. It is almost all second-growth forest because no old growth is
left. But it is designed to grow back into mature and old-growth status and, once it does, to sustain diversity properly. His model: the
design invented by conservation biologist Larry Harris, with help
from Art McKee, Jerry Franklin, Chris Maser, and others during his
sabbatical year in the Andrews.

Back in the Northwest, concepts developed at the Andrews, combined with key biodiversity protection approaches developed by conservation biologists, are finding their way even onto industrial lands.
In Washington and Montana, one of the nation's largest timber companies, Plum Creek, has adopted New Forestry wholesale. Once
called the "Darth Vader" of its industry by a Washington congress-
man, Plum Creek (once part of the Burlington Northern Railroad)
owns vast tracts of mile-square "checkerboard" lands, interspersed
with mile-square national forestlands. Burlington Northern acquired
the lands as national payment for driving rail lines westward in the
nineteenth century, and until recently its modern approach to logging has been to shave each square mile clean (making the less
aggressively harvested national forestlands, with their forty-acre cuts,
look wild in comparison). In addition to kinder and gentler forestry,
adopting logging schemes designed by Franklin, Plum Creek recently
signed a Habitat Conservation Plan with the U.S. Fish and Wildlife
Service, in which it will protect, beyond the bounds of the law,
unusually wide streamside forest buffers and roosting and foraging habitats for owls, and will provide habitats for a wide range of
imperiled animals, including some that probably have yet to return to
the area: wolves and grizzly bears. (It gets something major in

return: a controversial no-surprises guarantee for up to one hundred years that it can log the portions of its holdings the Fish and Wildlife Service has agreed are acceptable to log, without any future interference based on the Endangered Species Act.)

The Andrews group itself has entered a new phase in which it is actively trying to translate its theories about how these ecosystems work into large-scale active forest management. A large forested valley that lies directly beside the Andrews was designated in the Clinton Northwest Forest Plan as part one of ten large adaptive-management areas. There, the team has been conducting extremely detailed surveys to determine the history of fire in the large watershed. (They can look for characteristic, swirl-patterned, fire-generated ring scars on old stumps or in core samples of live trees. Fires occur at different frequencies in different parts of the landscape; some, for instance, are wetter and less susceptible. Nevertheless, almost all of the system is known to be driven—and succession at least partially "restarted"—by periodic disturbance, mostly in the form of fire.) Now that they have the fire scheme in place, they intend to develop a long-rotation logging scheme that will follow (somewhat) natural fire patterns. However, project leader John Cissel acknowledges that they do not intend to allow stands of trees to live as long as they might in nature. Cissel says the team members are assuming they can compromise: logging somewhat more frequently in many areas than nature might disturb, producing more wood fiber than they could by strictly following a natural rotation, but at the same time maintaining a more old-growth-like landscape that will sustain native species and a natural ecosystem. Only time and experimentation will tell. But at last they have a major hypothesis-based experiment to run, and perhaps Swanson's "physics envy" will one day find some real satisfaction.

But in an effort that surely would please the greenest among us more, an outfit called the Institute for Sustainable Forestry, and another called the Trees Foundation, in 1997 released a forest protection and logging plan for the sixty-thousand-acre Headwaters Forest, which includes some of the last and most spectacular stands of soaring redwoods in northern California. A company called Pacific Lumber owns Headwaters. Now owned by parent company Maxxam Corporation, under the sole control of a notorious corporate raider (who is presently in profound trouble with the federal government over his savings and loan dealings), Pacific Lumber once was a model corporate steward of forests, a careful, slow, and selective cutter. Since Maxxam raided Pacific's stock, it has been liquidating its corporate assets (redwoods) at a rate, critics charge, fantastically beyond any hope of regeneration. The apparent plan: rape and run. To put it mildly, it all has led to a pitched environmental battle. This year seventy-five hundred acres, important parcels of pure old growth, were bought in the public trust by the U.S. Congress. In the Institute for Sustainable Forestry plan, that group of environmental activist ecoforesters proposed that the balance of the lands be put on a very slow rotation management scheme, based on natural disturbance cycles. They aim to protect or develop a landscape characterized by large core areas of old growth, connected by corridors, "legacies," in the form of standing old-growth trees and logs, and six-hundred-foot streamside reserves, with particular attention to maintaining habitat for a wide array of species, including the predatory insects that will help naturally to control herbivory on trees.

Given both the pace and the significance of what a relative handful of scientists has discovered here, it seems undeniable that establishing sites, and teams of scientists, to study other kinds of ecosystems should be a major national, or global, priority.

In an elegant analysis of what he terms the invisible present, fisheries ecologist John J. Magnuson of the University of Wisconsin once offered a biological version of T. S. Eliot's notion of a world ending with a whimper. "Although serious accidents in an instant of human misjudgment can be envisioned that might cause the end of Spaceship Earth," he wrote in *Bioscience*, "destruction is even more likely to occur at a ponderous pace in the secrecy of the invisible present."

By way of explanation, he points out that all of us are capable of sensing changes in the environment over various time scales—from the very short ("the reddening sky with dawn's new light") to some changes that occur over longer stretches of time, such as the steady increase in wave strength on a lake as a thunderstorm gathers energy, or the seasonal changes in plants, or even time scales that we might consider long: the sense that fishing in a given lake had fallen off since one was a child, perhaps. But at time scales of decades or more, he says, we "are inclined to think of the world as static, and we typically underestimate the degree of change that does occur."

The "invisible present" describes slow changes that occur over decades, or across multiple generations, that simply cannot be seen by looking at a small slice of time. The changes wrought by synthetic chemicals in the environment could be such an event, or the slow but steady conversion of a forested rural area into a highly fragmented residential suburb, or, certainly, the conversion of a large wild forest landscape into a network of managed tree plantations. Other sorts of invisibility are more subtle. One example: throughout much of the western part of its range in Wisconsin and northern Michigan, one can find towering stands of hemlock trees, pieces of a forest ecosystem that looks to any eye like a parcel of wooded nature that is thriving and intact. But in fact, beneath most hemlock stands there is virtually no regeneration of young hemlocks, a problem that researchers have tied to the fact that deer populations have soared in recent decades to the point that the region's forests support more deer now than they did when the first white settlers arrived. (That in turn is a result of the fact that much of the forest before white settlement was old growth, which provides less ideal forage for deer than the younger forests that now dominate the region.) Hemlock are managing to produce new seedlings, but the seedlings often cannot survive intense browsing pressure from deer, for in winter, hemlock shoots thrusting through snows serve as highly favored foods—"ice-cream plants"—for deer, and thus seldom survive to maturity. In their version of the invisible present, the apparently thriving remnant stands of big hemlock are really "the living dead," as biologist Stephen Solheim once put it: when they inevitably age and die, no young hemlock will replace them.

It is only, Magnuson contends, through sustained, long-term research that this invisible present can be made visible, "much like time-lapse photography reveals the blooming of a flower or the movement of a snail."

He offers an exquisite example. At Lake Mendota, along which the University of Wisconsin lies, observers have been recording the duration of ice cover for the past 132 years. He points out that looking at a bit of data for a single year—he picks the winter of 1982–83—reveals nothing of any real consequence. But if ten years of data had been collected up until then, they would show a striking variation: ice cover in that one winter was forty days shorter than average for those ten years. If fifty years of data had existed, a clever scientist could see that such short-ice years occurred periodically and, it turns out, in perfect synchrony with a particular change in a certain climatic oscillation index thousands of miles away in the South Pacific: the phenomenon that has become so well known as El Niño. Suddenly, the "invisible" signal that could not be seen in data from a single year, or even from ten years, is now manifest, linking the patterns of ice on a Wisconsin lake to an important global climate phenomenon. By taking an even longer view, using the entire 132-year record of ice cover, an even more important and dramatic picture emerges: a clear, and statistically significant, signal from the ice showing another climate trend—a "general warming," albeit one occurring in fits and starts, over the entire period of time. The ice on the lake, in other words, confirms that as carbon dioxide emissions have increased over the past century and more, Madison, Wisconsin, has experienced a warming trend. From the years of patient data collection is emerging an important clue in the global warming puzzle.

In an accompanying article, the Andrews team's Swanson and Richard Sparks, a researcher at the Illinois Natural History Survey, seize on Magnuson's concept to suggest that ecosystems can also operate in "the invisible place." Research conducted on a limited geographic scale also risks similar failures, they suggest.

One simple example: a survey limited to looking only at a replanted forest on a cool, moist north slope of a mountain might conclude that all forests in an area are thriving, even while a more extensive look would show that regeneration on a hot, parched south slope has utterly failed. On a broader scale, a study using the sweeping geographic capabilities of a technology such as Warren Cohen's remote-sensing surveys can allow researchers to examine environmental conditions across an even greater range.

This book began by noting that almost all funding and incentive for biological—and even ecological—research worldwide has focused in the opposite direction: toward the short term and the small scale.

"For a variety of reasons," writes British ecologist Robert May, "many of our universities and other institutions do not easily accommodate work that stretches across traditional disciplinary boundaries, or that involves gathering data over a long time or a large area. For one thing, although departmental and other organizational boundaries are themselves usually the result of past evolutionary accidents, they are too often seen as absolute and inevitable; this can hinder new initiatives. For another thing, the time constraints of Ph.D. theses or the funding cycles of research grants understandably militate against long-term studies."

But he added that "too many ecologists have yielded to the temptation of finding a problem that can be studied on a conveniently small spatial and temporal scale, rather than striving first to identify the important problems, and then to ask what is the appropriate spatial scale on which to study them (and how to do this if the scale is large)."

In other words, a planet that Leopold characterized as a "world of wounds" desperately needs long-term, large-scale, interdisciplinary ecological studies. Simultaneously, it cannot easily have them because of the structures, and strictures, of the modern culture of science.

And there are more strictures. As much as any research program in the world, the Andrews effort has sailed against the tide, operating long-term, over large scales, and across disciplines. And yet repeatedly the research team ran into obstacles from scientific review panels questioning whether simply observing and recording natural phenomena is "real science."

"I spent a lot of my time in the early years just recording things like the date of leaf fall, or bud-break, or the time of flowering in various species," says Art McKee. "There's a lot of straightforward natural history kind of science that you have to do before you can even begin to ask questions—to propose worthwhile hypotheses. It's the first step in the scientific method. But if you try to get funding for that kind of work, the typical reply is: 'This is a descriptive study, and therefore unworthy of funding.'"

Ecologist May is nearly contemptuous of the notion that simply observing nature, and recording the relevant data, is somehow not important or "real" science. He notes that "current fashions" in science rise largely from the work of Sir Karl Popper, who proposed in 1959 the notion that the primary characteristic of good science was a "falsifiable hypothesis," one that can be refuted through a clear and repeatable test. (Simplified, this "null hypothesis" approach works like this: A scientist who believes a dropped stone must fall to earth will propose a hypothesis suggesting the opposite—the stone will *not* fall to earth. A simple, repeatable experiment—dropping a rock— will disprove the null hypothesis.) As science historian Peter J. Bowler has pointed out, the approach works ideally in the realm of chemistry or, especially, physics, but not always so well in either geology or ecology, where "to describe and classify the entities that make up our environment, and to explain how those entities were created by natural process, involves the construction of explanatory schemes."

When it comes to ecology, Robert May suggests, Popper's notions "owe more to philosophical musing than any real appreciation of how physical scientists actually work." Nevertheless, as Bowler points out, an obsession with hypothesis-testing-based science has led some physicists and chemists to question whether disciplines like geology and ecology can be called sciences at all. At the very least, he points out, they have often been characterized as "soft" sciences.

Given the constraints from the culture of science, it might seem astonishing that the Andrews project has been able to succeed at all. The better news is that it was not entirely alone in its mission even in its earliest days, but one of a cluster of struggling whole-ecosystem programs that finally began to convince the National Science Foundation—the country's leading funding source for scientific research in the United States—of the worth of such research in general. (Another of those early model projects—begun even sooner than the Andrews effort—deserves some mention here. In New Hampshire, the Hubbard Brook ecosystem study, also in a Forest Service experimental forest, broke early ground with electrifying research on nutrient flow in forest systems. In that ecosystem, a once cut-over but regenerated second-growth forest, researchers Gene Likens and Herbert Bormann first determined that virtually all of the nutrients that entered the experimental forest, whether through leaf fall or rainfall, remained bound up in the ecosystem, continually recycled and retained by living plants and animals. After that painstaking data collection was complete, the researchers had one of the forest's six valleys clear-cut. Nutrients promptly began to surge out of the stream that drained the valley: nitrate loss alone increased sixtyfold. It was here, too, that Likens and Bormann first pieced together data about the changing chemistry of the precipitation falling on the ecosystem, and thus first raised the alarm about acid rain falling on the United States.)

In 1980, the National Science Foundation established what it termed its Long-Term Ecological Research (LTER) Program. The program began with the Andrews, Hubbard Brook, and five other sites. Today in the United States there are twenty-one sites scattered from the Alaskan tundra to the deserts in New Mexico to the rain forests of Puerto Rico, under the LTER umbrella.

After nearly two decades, the LTER effort in many ways remains in its infancy: at not even one of the sites do researchers have a complete catalog of all the species present. (Vertebrate animals and plants typically have been counted and mapped, but that's a far cry from cataloging the usual myriad of insects, fungi, algae, molds, bacteria.)

Still, just as with the Andrews, intriguing discoveries have already emerged. Wisconsin's John Magnuson, who serves as Swanson's counterpart as leader of the North Temperate Lakes LTER project in that state, and his research team have found that a lake's chemistry is intimately linked to subtle differences in its relative elevation and to the vagaries of time. In lake-riddled regions where lakes flow into each other not through channels, but unseen, through the soil, water can take years to move only about a hundred yards from one lake to the next, and a century or more to move to the lowest lakes on a landscape. Along the way, the moving water dissolves minerals, which increases its alkalinity as it moves from lake to lake. Thus, a savvy scientist can predict simply from elevation maps which lakes will be relatively more sensitive to problems such as acid rain, and which will be better-buffered by their store of acid-neutralizing alkalinity.

One discovery at the Sevilleta Desert LTER site in New Mexico demonstrates the power of whole-ecosystem research in terms of human ecology, even human life. In 1993 a mysterious and fatal respiratory disease broke out on and around the vast Navajo reservation in New Mexico. Medical investigators eventually determined the disease was caused by normally rare hantavirus. Once they had linked the virus to fecal dust from infected deer mice, they began to wonder if they were staring down the barrel of a new and devastating epidemic. Why, after all, had this rare virus suddenly appeared in such force? Researchers at the Sevilleta LTER offered an answer: it was no long-term epidemic, only a consequence of, well, a certain climatic "oscillation" hundreds of miles away in the Pacific. In that year, an El Niño weather pattern once again dominated the eastern Pacific Ocean. One result had been a wet year in the southwestern deserts. The rain had triggered an explosion of seed-bearing plants. The seed abundance had, in turn, triggered an explosion in small seed-eating mammals, including the deer mouse. The mouse population and the hantavirus infection rate would plummet as the El Niño dissipated, the researchers told worried public health officials, but the problem could arise again with the next cyclical El Niño. The next year, the weather system dissipated, and the hantavirus problem vanished. (In 1998 another El Niño warmed the eastern Pacific. That June, *New York Times* reporter James Brooke wrote, "Carried by a surging population of deer mice, hantavirus has returned to the Southwest this spring. . . .")

That sort of discovery, says Fred Swanson, is part of the excitement of ecosystem research. "It's a frontier issue," he says. "People are still out there on the scientific frontier, looking in new places, and finding interesting stuff." But the real power of the approach, he says, is tangible and utterly practical. "These are the ecosystems that feed us, provide our drinking water, give us aesthetic pleasure." In light of that, he says, "the time and the resources devoted to these studies is a pretty small price to pay."

Perhaps too small. If anything, the pace and the wonder of discovery in the Andrews suggests just how little we know about the way the earth's natural systems function, and how important it is that we learn more. For all the positive aspects of a nationwide LTER program with twenty-one sites, its greatest drawback might be that it comprises *only* twenty-one sites. There are no whole-ecosystem study programs in even some of the ecoregions considered unique and most valuable in the United States: the fantastically biodiverse sage-scrub chaparral systems of southern California (where real estate developers are building suburbs and malls at a breakneck pace), the spectacular mixed mesophytic hardwood forests of the Cumberland Plateau (where a series of new giant wood-chip factories have recently begun quite literally to devour a forest ecosystem that has a single parallel on earth: and that one in central China), or the northern hardwood and transitional forests of the Great Lakes (where the woods is in recovery from a rapacious bout of logging at the turn of the century but where the attention of timber companies is already turning, in great force, next). No one in any of these places is attempting the kind of detailed, interdisciplinary, and long-term studies that would allow, say, another team of entomologists to begin to understand how insects might drive the life of trees, or truffles, or vice versa.

Given the critical importance of understanding more about how life on earth exists and survives, present funding for the Long-Term Ecological Research Program—about 12 million dollars in fiscal 1999—seems not much more than chicken feed compared to, say, programs such as defense research or the exploration of space. In a 1990 report, conservation biologist Jared Diamond noted that the United States had spent about one billion dollars on the first Viking mission to Mars, partly to find out if any signs of life existed there, and noted that this sum dwarfed funding for exploration of life on Earth.

Perhaps the lack of funding to expand and improve ecosystem research is not so much a matter of too little money, but too little attention to priorities. What if we were simply to wait—a decade, a few decades, even a century or more—to proceed with the exploration of space, and, for a time, reallocate those billions of dollars to in-depth studies of the earth's vastly unknown living systems? As I write these words, a national news network has reported that, for want of about seventy million dollars, a critical module for a new international space station lies nearly complete, but not complete enough to launch, in Russia. Russia's economy is sliding into a steep recession, and its government says it is too broke to complete the module. In a fit of pique over escalating Russian costs and declining efficiency, the U.S. Congress refuses to provide funds to help complete the module. Meanwhile the U.S. facilities and personnel are marking time, waiting for the module, at an estimated cost of more than one hundred million dollars each month! The investment interest from three months' worth of this wasted money could run a major whole-ecosystem site the size of the Andrews for a century.

Eminent scientists have long called for a major boost to ecosystem research and its funding. If a quarter century of work in the Andrews has proven anything, it is that new opportunities for discovery abound in hidden folds of the earth's ecosystems. In the end, can there be anything more extraordinary or bizarre or perverse than a culture that pours its resources into the study of the mystery of other worlds, even as it fails to take far more modest steps to uncover the interwoven wonders of life on its own?

Selected Readings

Some of these books and articles directly informed research for this book, others provided related reading. I relied particularly on Worster, Bowler, and Egerton for information on the history of ecology as a science, on Williams and Cox and others for background on the history of forestry in America, and on Durbin and Ervin for history of the spotted owl controversy. For readers interested in more technical resources about the work of the Andrews group, contact the Pacific Northwest Research Station, U.S. Forest Service, Corvallis, OR 97311, for their catalog, *Research Publications of the H. J. Andrews Experimental Forest*, including periodic supplements. It is also available at http://www.fsl.orst.edu/lterhome.html.

Allen, Timothy F. H., and Thomas W. Heokstra. *Toward a Unified Ecology.* New York: Columbia University Press, 1992.

Alverson, William S., Walter Kuhlmann, and Donald M. Waller. *Wild Forests, Conservation Biology and Public Policy.* Covelo, Calif.: Island Press, 1994.

Bacig, Tom, and Fred Thompson. *Tall Timber: A Pictorial History of*

Logging in the Upper Midwest. Bloomington, Minn.: Voyageur Press, n.d.

Botkin, Daniel B. *Discordant Harmonies: A New Ecology for the Twenty-First Century.* New York: Oxford University Press, 1990.

Bowler, Peter. *The Norton History of the Environmental Sciences.* New York: W. W. Norton, 1992.

Carroll, George. "Fungal Endophytes in Stems and Leaves: From Latent Pathogen to Mutualistic Symbiont." *Ecology* 69(1) (1988): 2–9.

Commoner, Barry. *The Closing Circle.* New York: Knopf, 1971.

Conner, E. J. H. *The Life of Plants.* Chicago: Chicago University Press, 1964.

Cox, Thomas R., Robert S. Maxwell, Phillip Drennon Thomas, and Joseph J. Malone. *This Well-Wooded Land: Americans and Their Forests from Colonial Times to the Present.* Lincoln: University of Nebraska Press, 1985.

Denison, William C. "Life in Tall Trees." *Scientific American* 228(6) (1993): 74–80.

Durbin, Kathie. *Treehuggers.* Seattle: The Mountaineers, 1996.

Egerton, Frank N. "Ecological Studies and Observations before 1900." In Benjamin Taylor and Thurman White, eds., *Issues and Ideas in America.* Norman: University of Oklahoma Press, 1976.

Erlich, Paul, and Anne Erlich. *Extinction.* New York: Ballantine, 1981.

Ervin, Keith. *Fragile Majesty.* Seattle: The Mountaineers, 1989.

Franklin, Jerry. "Toward a New Forestry." *American Forests* (November–December 1989).

Franklin, Jerry, Kermit Cromack, Jr., William Denison, Arthur McKee, Chris Maser, James Sedell, Fred Swanson, and Glenn Juday. *Ecological Characteristics of Old-Growth Douglas Fir Forests.*

Corvallis, Ore.: Pacific Northwest Forest and Range Experiment Station, 1981.

Harris, Larry D. *The Fragmented Forest: Island Biogeography Theory and the Preservation of Biotic Diversity.* Chicago: University of Chicago Press, 1984.

Harrison, Robert Pogue. *Forests: The Shadow of Civilization.* Chicago: University of Chicago Press, 1992.

Kohm, Kathryn A., and Jerry F. Franklin. *Creating a Forestry for the Twenty-first Century: The Science of Ecosystem Management.* Covelo, Calif.: Island Press, 1997.

Leopold, Aldo. *A Sand County Almanac (With Essays from Round River).* New York: Ballantine, 1966.

MacCleery, Douglas W. *American Forests: A History of Resiliency and Recovery.* Washington, D.C.: USDA Forest Service, 1992.

Magnuson, John R. "Long-Term Ecological Research and the Invisible Present." *Bioscience* (July–August 1990): 495–501.

Maser, Chris. *Forest Primeval.* San Francisco: Sierra Club Books, 1989.

Maser, Chris, and James M. Trappe, eds. *The Seen and Unseen World of the Fallen Tree.* Corvallis, Ore.: Pacific Northwest Forest and Range Experiment Station, 1984.

May, Robert M. "The Effects of Spatial Scale on Ecological Questions and Answers." In P. J. Edward, R. M. May, and N. Webb, eds., *Large Scale Ecology and Conservation Biology.* Boston: Blackwell Scientific Publications, 1993.

Myers, Norman. *The Sinking Ark.* New York: Pergamon Press, 1969.

Norse, Elliot A. *Ancient Forests of the Pacific Northwest.* Covelo, Calif.: Island Press, 1990.

O'Toole, Randall. *Reforming the Forest Service.* Colvelo, Calif.: Island Press, 1987.

Perlin, John. *A Forest Journey: The Role of Wood in the Development of Civilization.* New York: W. W. Norton, 1989.

Perry, D. A., M. P. Amaranthus, J. G. Borchers, S. L. Borchers, and R. E. Brainerd. "Bootstrapping in Ecosystems." *Bioscience* (April 1989): 230–237.

Pinchot, Gifford. *The Fight for Conservation.* Seattle: University of Washington Press, 1973.

Platt, Rutherford. *The Great American Forest.* Englewood Cliffs, N.J.: Prentice-Hall, 1965.

Pyne, Stephen J. *Fire in America: A Cultural History of Wildland and Rural Fire.* Princeton, N.J.: Princeton University Press, 1982.

Raphael, Ray. *Tree Talk: The People and Politics of Timber.* Covelo, Calif.: Island Press, 1981.

Richards, John F., and Richard P. Tucker, eds. *World Deforestation in the Twentieth Century.* Durham, N.C.: Duke University Press, 1988.

Robinson, Gordon. *The Forest and the Trees: A Guide to Excellent Forestry.* Covelo, Calif.: Island Press, 1988.

Sedell, James R., Jerry F. Franklin, and Frederick J. Swanson. "Out of the Ash." *American Forests* (October 1980).

Siedeman, David. *Showdown at Opal Creek: The Battle for America's Last Wilderness.* New York: Carroll and Graf, 1993.

Swanson, Frederick J., T. K. Kratz, N. Caine, and R. G. Woodmansee. "Landform Effect on Ecosystem Patterns and Processes." *Bioscience* (February 1988): 92–98.

Swanson, Frederick J., and Richard E. Sparks. "Long-Term Ecological Research and the Invisible Place." *Bioscience* (July–August 1990): 502–508.

Thoreau, Henry David. *The Maine Woods.* Princeton, N.J.: Princeton University Press, 1972.

Wilderness Society. *America's Vanishing Rain Forest.* Washington, D.C.: The Wilderness Society, 1986.

Williams, Michael. *Americans and Their Forests: A Historical Geography.* New York: Oxford University Press, 1989.

Wilson, B. F. *The Growing Tree.* Amherst: University of Massachusetts Press, 1970.

Worster, Donald. *Nature's Economy.* Cambridge, England: Cambridge University Press, 1977.

Xerces Society. *Wings: Essays on Invertebrate Conservation* (Summer 1990). (This colorful issue includes essays by the Andrews's Jack Lattin and Andrew Moldenke on insects and other arthropods in the forest.)

Zazlowsky, Dyan, and the Wilderness Society. *These American Lands.* New York: Henry Holt, 1986.

Index